IN THE
HERDER'S BOOTS

Stephen Parliament

IN THE
HERDER'S BOOTS

Challenging Life of the
Nomadic Cashmere Goat
Herder in the Gobi Desert of
Mongolia

Asian Studies

Collection Editor
Yongtao Du

LPp

To Laurie, whose love and support never wavered over time and miles.

First published in 2025 by Lived Places Publishing

The author and editor have made every effort to ensure the accuracy of the information contained in this publication, but assume no responsibility for any errors, inaccuracies, inconsistencies, or omissions. Likewise, every effort has been made to contact copyright holders. If any copyright material has been reproduced unwittingly and without permission the Publisher will gladly receive information enabling them to rectify any error or omission in subsequent editions.

British Library Cataloguing in Publication Data
A CIP record for this book is available from the British Library

ISBN: 9781917566094 (pbk)
ISBN: 9781917566117 (ePDF)
ISBN: 9781917566100 (ePUB)

The right of Stephen Parliament to be identified as the Author of this work has been asserted by them in accordance with the Copyright, Design and Patents Act 1988.

Cover design by Fiachra McCarthy
Book design by Rachel Trolove of Twin Trail Design
Typeset by Newgen Publishing UK

Lived Places Publishing
P.O. Box 1845
47 Echo Avenue
Miller Place, NY 11764

www.livedplacespublishing.com

Abstract

Traveling throughout the Gobi Desert of Mongolia, living with nomadic cashmere goat herders, and helping them organize marketing cooperatives to get a fair price for their fantastic fiber gives me respect for these beautiful people, their determination, and an appreciation for their pride. They are now seeking a place of economic security while surrounded by voracious neighbors of China to the south and Russia to the north, not a situation that instills peace of mind. This is the story of my small role in living and working with them as they maneuver into the modern world after Soviet domination.

Key words

cashmere, Gobi Desert, herder, Mongolia, nomad, Ulaanbaatar, U.S. Agency for International Development (USAID)

A note on terminology

AMAG	Also 'aymag', 21 provinces forming the first administrative subdivisions of the country; Ulaanbaatar is the 22nd province, separately governed.
CULL	To slaughter an animal to reduce herd size or to eliminate an undesirable genetic characteristic in that animal, such as weak bone structure, jaw overbite, or too much coarse hair.
CHANGE PEOPLE OR CHANGERS	Agents, or "changers", facilitate exchange between herders and processors in the fiber, hide, and meat markets in UB, extracting commissions in the transaction, according to some sources, predominantly Chinese.

DEL or DE'EL (děl)	Quilted winter garments, usually in dark blue or maroon, with a bright orange or yellow sash and extra-long sleeves that extend past the fingertips for warmth and to eliminate the need for gloves. Worn by both men and women.

Figure 1 Our consultant, Kris, is trying on a new deal or "de'el" that is being sown by Chuulunbaatar's wife, Udval, with the assistance of my colleague Alta.

DZUD OR ZUD	An extreme weather condition resulted in animal starvation and death. Some zud are called 'white' or snowy. In such cases, animals cannot find grass under the snow; some dzuds are due to extreme cold with a lack of snowfall and subzero temperatures resulting in draught the following spring.
HERDER	The first and most striking quality of Mongolian society, from the stories of Batsuh, Alta, Dr. Oyun, and others, is women's powerful role in sharing work, being educated, and keeping the country moving. The word "herder" is not "herdsman" because many of the herders and women and children."
MOBILE PASTORALISM	The pattern among rural cashmere goat, sheep, camel, and other domestic animal raisers in arid climates in which herders must constantly move herds to take advantage of better grazing opportunities, often connected with other related economic activities such as processing animal products, selling crafts, and bartering for staples.

Figure 2 Chuulunbaatar's son Batzorig, with goats searching for sparse grass in the arid desert

NOMADISM	The term refers to a wandering culture in which people and their animals constantly migrate to exploit grazing opportunities (mobile pastoralism) or trade opportunities. They secure goods when available and move them to markets as opportunities arise.

SUM	(Pronounced "soom") Administrative subdivisions of each province or "aimag" providing direct services. There are 331 sums in 21 aimags that comprise the country. A "sum center" is the commercial town or village of the sum.
UB	Abbreviation for Ulaanbaatar, the 22nd aimag administrative unit, comprises the capital city. Also, 'Ulan Bator'.

Contents

Preface/ Introduction

The New Millennium

I am entranced, feeling otherworldly, and delighted at having the good fortune to be here in Ulaanbaatar, the capital city of Mongolia.

Figure 3 Ulaanbaatar, capital of Mongolia with a statue of Genghis Khan

My title is Program Manager for Agriculture with Mercy Corps. Of course, I was excited when the United States Agency for International Development (USAID) asked me to join a team already in the country that needed a specialist in cooperative marketing. That is my background, so here I am.

After the initial thrill of seeing goats grazing just outside the airport terminal, the self-doubts emerged: Do I honestly have anything to offer this ancient society of nomadic herders that they have not already figured out after thousands of years of survival in this brutally cold, arid climate and having built one of the largest empires in the history of the world?

I shall gladly impart anything useful I know. I believe in the principles of cooperative organization and that they are universal. We shall see.

Mongolia is an independent and democratic state, having emerged in 1990 from seventy years of Soviet domination. The Mongolian Empire was once the largest in the world, stretching from Eastern Europe to Central Asia to the east coast of Siberia and China. It is now an enclave of a proud culture, with over six thousand miles of unprotected borders with Russia to the north and China to the south, both countries with voracious appetites for expansion. As a rare democracy and free market economy of all the previous Soviet satellites, they are a natural ally of the United States, which has taken a focused interest in their survival.

However, the survival challenges are complex, environmental, political, and economic. The people possess significant natural resources, spectacular animals, and the world's finest cashmere fiber and are endowed with determination and ingenuity. They rejected Soviet collectivism and were curious about how to participate in the world of international trade, or they would not be so open and receptive to the overtures of USAID. From what I learned about the country before arriving, the foundation of experiences and values will be compatible with ideas of cooperating in a free market instead of a demand economy.

This is the story of the struggle to survive in one of the harshest physical environments in the world. How an entire country shares the "commons" to their mutual benefit is a model for the world.

This book emerged from my field notes, and subsequent interviews with Mongolian friends and colleagues committed to the autonomy and development of their country. As I traveled the Gobi Desert with my Mongolian colleagues, I kept daily notes because I knew something unique and irreplaceable was happening. This transition from Soviet collectivism to independent cooperation in the whole country was an experiment not easily duplicated in world history. The country communally owns all the land. No private property, fences, or "No Trespassing" signs exist.

In the early spring, the herders get together and demarcate the general areas where each family will circulate with their animals, looking for grass but moving so as not to overgraze or run into each other.

The facilitator of this venture and the organization I worked for is Mercy Corps International, a major nonprofit provider of services to USAID. The title 'Gobi Initiative' is the trade name for our specific project. We were funded by a $24 million five-year grant from USAID. Mercy Corps applied for the grant competitively against several other service and academic organizations. I was initially a team member organized by Professor Judith Gillespie, a friend and colleague from graduate school at the University of Minnesota and then at the State University of New York, Albany. I am indebted to her original invitation and my introduction to Mongolia.

Our proposal was a serious contender for the grant and was one of three finalists. In preparing the application, our group was awarded $15,000 to travel to Mongolia, meet people, visit various development organizations, and travel in the countryside, talking with herder families who might eventually be interested in participating in our program if we were successful. Our team was putting our application together in Mongolia for about a month. The USAID staff said that they always reserve the right to keep people interested in international development in mind. They may contact individual team members to join a successful applicant later if an individual has a skill that the contracting team needs. That is what happened in my case.

After returning from Mongolia and preparing the final application, I was contacted by the group that was ultimately successful: Mercy Corps and the Gobi Initiative. They asked if I was still interested in working in Mongolia and, if so, if I would interview for an opening they had on their staff. The work would be in agricultural development, specifically in the formation of marketing and production cooperatives, which is an area in which I have professional expertise. I was thrilled, having long since abandoned the initial excitement of going to such a remote and unfamiliar setting. I accepted, interviewed, and left for the Mercy Corps headquarters in Portland and then to the capital of Mongolia, Ulaanbaatar, one month later. (I will frequently use the abbreviation "UB" for Ulaanbaatar because that is what most Mongolians use in conversation). I put all my personal belongings in storage, closed up the housing development and property management contracts I had in Minneapolis, packed one huge duffle bag, and sold some stuff.

The most challenging part of leaving, however exciting, was telling a beautiful lady with whom I had just fallen in love that I cherished our new and blooming relationship but that I was going to Mongolia, maybe for as long as five years! She had two teenage sons, a house, a dog, and a teaching career, so she was not likely to join me. I assured her she could visit. Somehow, it did not grab her like it did me. But don't fear. It is not forsaken. She will reappear at the end of the story and periodically in the middle. She was more than supportive and turned out to be an amazing correspondent on the new medium of email that I was soon to master. Our fledgling relationship somehow flourished through electronic communications. She is now my wife.

I arrived in UB in the fall of 1999, nine short years into independence, and the country was struggling with a bitterly cold winter, followed by summer drought and burning hot temperatures. In the Gobi Desert, this weather cycle called a dzud in Mongolian, produces very little grass. To the herders, it means the death of millions of animals.

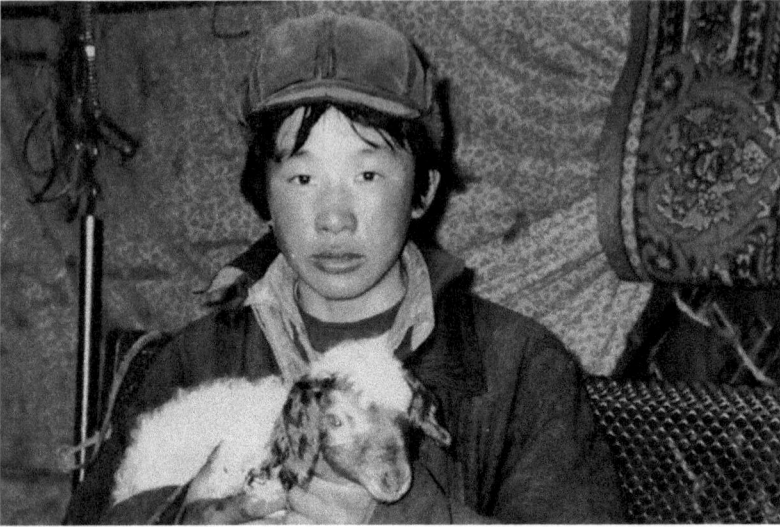

Figure 4 On brutally cold winter nights, a young herder brings newborns into the family ger at night to keep them warm.

The organization I joined, the 'Gobi Initiative', worked with nomadic cashmere goat and camel herders in the Gobi Desert. I more than fell in love with the solemnity and dignity of the people, the integral connection among humans, animals, the land, and nature in surviving one of the harshest environments on earth, and the rare privilege of living within one of the oldest and once most powerful civilizations in human history.

The story of the indomitable Mongol is not well known in other parts of the world. This is because their resources are in the cross-hairs of resource-hungry countries on their immediate northern (Siberian Russia) and southern (China) borders and are already being exploited, either directly by mining companies or indirectly by investment from every other developed country.

The Mongolia Society at Indiana University published a factual account of my work[1], but that article did not capture the emotive connection to the people. I continued to stay in touch with my friends and contacts in Mongolia. In the winter of 2009, news accounts described the disappearance of hundreds of streams and lakes in southern Mongolia and the devastation of pastoral life caused by overgrazing. Louisa Lim of National Public Radio[2] recently presented a week-long account of life in Mongolia in which she described how some herders in Ömnögovi' aimag in the southeast Gobi Desert are being forced to give up herding because of competition for grazing land from the mining interests. Some of these herders have tried gold mining independently but have been run off by mining company security guards. The mining process also uses significant amounts of water, causing the depletion of the already scarce streams and lakes critical to herding. In the December 8, 2009 Business of Green section of *The New York Times*, Sarah J. Wachter[3] describes the "unraveling of pastoralism", in which overgrazing contributes to the destruction of the fragile grazing land in the Gobi. This phenomenon started with the *dzud* in 1999 and continues today. We may watch the end of a culture as independent nomadic families move with their animals, sharing less and less productive land with other herders in a random pattern of searching and pasturing. No other place on earth organizes animal husbandry this way.

As I write these notes and reflections more formally, I know this book's timing is appropriate. When I recently heard about President Putin staging a "friendly visit" to Mongolia, I could sense the wolf at the door. My attention spikes, and friends say I need to write about the place because it is a mystery to most people. We need to pay attention. Just ask the Ukrainians.

It was time, I decided, to contact my former colleague Alta.

Figure 5 Batzorig (first son), Chuulunbaatar, and Alta checking herd inventory.

I called Mrs. B. Altantsetseg (Alta), my co-program Director for Agriculture at the Gobi Initiative, to ask about her assessment of the dzud and recent accounts I had heard of the devastation of European and Canadian mining operations in the Gobi.

She answered her cell phone while in the hospital, pregnant and with high blood pressure.

"Alta, I didn't know you were in the hospital. I don't have to talk to you now. Why did you answer the phone?"

"Why not? I can talk," she said. "Just because I am in the hospital with high blood pressure and about to give birth any minute doesn't mean I can't talk."

She was working for the Millennium Challenge Fund, a multi-million-dollar development account set up by the government of Mongolia and funded by the United States. "The mining is good for jobs and investment but could harm the herders. They take grassland and water," she said. "Herders are doing many different types of businesses besides herding. The nomadic way of life will change," she said, "first because it should, but also because it is being pushed out. Some things must go ahead."

"Are we looking at the end of nomadism in Mongolia?" I asked Alta.

"The grass is disappearing—too many animals. The only way to control it is for cooperatives of herders to lease large sections of land and limit their membership. Then they will prosper," she said. "The herders have to control the land themselves. Cooperatives can decide how many herders can use the land. The government can't do that, only the herders. If we lease land to herder coops, they can protect their investment in drilling wells and setting fences," she said. That last word, fences, was a word I never expected to hear from her.

I interpreted her to mean that the nomadic life might be finished but that cashmere goats and herding can survive, only in a different form.

On October 6, 2009, the Prime Minister of Mongolia signed a development agreement with Ivanhoe Mines Mongolia, Inc. for extensive mining rights in the South Gobi. The agreement was made possible when the Mongolian Parliament removed a windfall profits tax on mining operations. As compensation to the people of the country, Dr. S. Oyun, a new member of the Mongolian Parliament and a former geologist for the international mining company Rio Tinto is one of the major investors

in the mining operations, inserted a provision into the development agreement for Mongolia to own 34per cent of Ivanhoe
Mines Mongolia, Inc. to protect the people's interests.

The lives of these herders and of the last nomadic civilization on
earth are changing rapidly, irrevocably, either in the direction of
cooperative growth or toward possible extinction. The people of
Mongolia sit on top of great wealth, a wealth they may never see,
buried just below the surface of the sand in the Gobi Desert. Their
country is in the shape of a walnut wedged between the giant
nutcracker of Siberian Russia and North China, with Canadian,
English, and Australian mining interests digging like fury before
the nut cracks.

The people in this narrative are ready to talk and want to, with
one exception. On November 13, 2008, *The New York Times*
reported: "U.S. Aid Worker Slain in Pakistan."[4] Stephen Vance
was our Chief of Party at the Gobi Initiative and my immediate supervisor. He was assassinated by Islamic terrorists in
Peshawar, Pakistan, where he was working for the Foundation
for Cooperative Housing International on a USAID contract. He
was trying to create economic opportunities through small business development when he was ambushed on his way to work.
While still in UB, he had married a Mongolian woman with five
children, who now live in the U.S. I am stunned and saddened by
his tragic death.

Alta is the Peri-Urban Property Rights Director at the Millennium
Challenge Account. She is working on land-lease arrangements
with herder cooperatives. This work concluded in 2013. The term
"peri-urban" is roughly equivalent to suburban or even exurban,

describing new settlements on the far outskirts of an urban area. In Ulaanbaatar, it means that herders are moving to the outer edges of the city, bringing their families, living quarters called "gers" or the more familiar Russian word "yurts", and some animals. Alta is trying to help them with cooperative arrangements so that several herders can work together to raise animals and take goods to the markets in UB.

An inspiration for this book comes from Dr. S. Oyun (or in some spellings, Oyuun), founder of the Civic Will Party and, more recently, leader of the Democratic Party of Mongolia and a Member of Parliament. She is also a supporter of the Women's Information and Research Center and the Gender Center for Sustainable Development. She is the sister of S. Zorig, who in 1998 was nominated by the Democratic Party as their candidate for Prime Minister. He had campaigned vigorously for independence from Soviet rule and for transparency in government. He opposed the secret deals that were being negotiated with Russian mining interests. A few hours after his nomination in 1998, he was assassinated. Dr. S. Oyun was a geologist in Cambridge, England, working for Rio Tinto, an international mining company based in London, which is now interested in investing in Ivanhoe Mines. She returned to Mongolia to run for her brother's seat in Parliament and is now a successful politician. She advocates for women's interests in all spheres of economic life, including herding and job creation through mining. She understands the conflicts facing the country and is available for an interview.

Finally, endless ideas and mental energy come from an old acquaintance from when I worked at the Gobi Initiative, a wonderfully committed person named Maidar. He and Alta, among many others I interviewed, represent the thoughtful future. Maidar is not very interested in the potential riches of mining operations. The future lies in the creation of Mongolian businesses that collaborate with herders, business people in UB, and government officials who are willing and able to help build new forms of cooperative enterprises.

It is for the herders I worked with at the Gobi Initiative – Chuulunbaatar, Dembereldorj, and Tömör – Alta and her family, and Maidar; as well as in memory of Stephen Vance that I write this book.

All of the opinions and interpretations in this manuscript are solely those of the author and are not intended to reflect the positions of the Government of Mongolia, the United States Agency for International Development, or the U.S. Department of State. Any inaccuracies are entirely the fault of the author, who, nonetheless, is deeply indebted to the people of Mongolia, the staff of the Gobi Initiative, and all the non-governmental organizations working to promote civil society and a healthy economy in Mongolia, which remains a free, independent and democratic state.

The book is richer by far because of the thoughtful contributions of "Alta, my colleague at the Gobi Initiative, who is the key character in the book; Degi Tserendamba, a key collaborator and research assistant who recently graduated from the University of Minnesota and is now with the Mongolian National Psychological Association in U.B., Tserenchunt Legden and her daughter Delgerjargal Uvsh; and my constant inspiration, Maidar, all of whom contributed in their own way to this project.

As carefully as possible, I kept track of the people referenced in my notes and tried to ensure correct spelling and affiliations. I apologize for any mistakes. Titles and organizational affiliations have undoubtedly changed since my initial contact. The information I collected through interviews and conversations is accurate to the best of my ability as a participant observer. A list of people cited in the book is in an appendix at the end.

I am indebted to Dr. Brian Baumann at Indiana University for his meticulous historical and linguistic editing; to The Mongolia Society at Indiana University and to Susie Drost, Executive Director, for continued advice and support; and to Marsha Stelzer, Kirsten Neuhaus, and Prof. Judith Gillespie.

For the germination of my notes into a book, I want to give special thanks to the writers' group of Spooner, Wisconsin: Eva Apelqvist, Judith Barisonzi, Joel Friederich, Bob Hasman, Kevin McMullin, and Denise Meister, all of whom provided encouragement and criticism of the best kind. I also extend my gratitude to colleagues and students at Wisconsin Indianhead Technical College, Rice Lake, and the College of Educational and Professional Studies, Department of Teacher Education, University of Wisconsin, River Falls who have provided useful comments and suggestions on the text.

Learning objectives

- Learn the values and operating principles needed for a country that was a satellite of the Soviet Union to emerge as a successful, independent, and democratic state;

- Define nomadism and give specific examples of how it is still a functioning system of agricultural and economic life in Mongolia;

- Use a center-periphery graphic model, to gain an understanding of how products and goods flow through the country and where the markets are located;

- Understand how the country can function when private farmers, businesses, or local units of government neither own nor control any land;

- Define and document three major obstacles to the integration of Mongolia into the modern world.

1
Where the cashmere trail begins: Search the Gobi Desert for the herder Chuulunbaatar's winter ger settlement and find nomads

Figure 6 Dusk in the Gobi

We are not lost, but we do not know where we are. The horizon and sky become the same curved plane in the Gobi Desert, so that your feet are on the same surface as the blank, white, cloudless sky, where your eyes search for a focal point, finding none. The 1,700 km of paved road dissipates into wider and wider ruts, then into meandering camel tracks and gullies crisscrossing old tire tracks from previous caravans.

An old trail can be a quarter of a mile wide and branch into two or three directions, seeming to search for a destination. In the Gobi Desert, families, their animals, and their domiciles, called gers (or the more familiar 'yurts' in Russian), are in constant motion.

The Gobi is just a word for dozens of plateaus and valleys in southern Mongolia between the Altai Mountains to the north and the Tibetan Plateau to the south, stretching eastward to the North China Plain. It can be brutally hot or penetratingly cold. Traveling today in the winter, the average low is 40°F below zero, while the average summer high is 113°F. In winter the way the wind feels on your skin is -60°F.

At the top of a mountain pass, our driver pulls the Land Rover to the side of the road and stops. He hops out and hurries over the ridge and out of sight to relieve himself.

Alta gets out of the Land Rover, stretches, and looks into the distance without focusing on anything.

It is how Mongolians look at the horizon and beyond , off the world's end. Alta is my Mongolian counterpart at the Gobi Initiative. We are co-Program Directors for Agriculture, a program run by Mercy Corps International, focusing on civil society and agricultural development in Mongolia. A third member of our party is Kris, an accomplished goat breeder and veterinarian from the United States. Our translator and guide is a quiet and thoughtful gentleman named Buyan, who was raised in this region and knows it by instinct. I get out and stand next to them, staring into the cold gray expanse of the Gobi, looking for a fresh trail, a herd of camels, or the curl of smoke rising from a settlement. With fewer than two people per square kilometer, there are only a few people within my range of sight, if they could be seen at all.

Figure 7 Cashmere Trail

Our destination is the winter settlement of the nomadic herder Chuulunbaatar and his family. They are wintering now, waiting for a spring thaw, but they do not necessarily winter in the same place every year. The choice of location for a nomad is solely a matter of available vegetation. Where they last wintered is no help to us. Even if they have been circling within a hundred square kilometer area for the past five years, the family may move a few hundred kilometers away to an entirely new area as the land becomes overgrazed.

"If we spot a herd of goats, we can find a settlement of gers," says Alta. "If it is not Chuulunbaatar, whoever lives there will know where he is."

Gers are the color of the desert. They are low circular lodges made from felt and hides that are easily rolled up and transported by camel when a nomadic family needs to move, which happens about eight to ten times a year when the goats can no longer find anything to eat. The distinguishing feature of the ger is the door, which is about four feet tall and painted in bright colors. The inside of the ger is warm and bright, with beautiful patterns painted on wooden slats, a shrine to family ancestors, and small pieces of furniture. Once inside, the significance of the Mongolian word "ger" becomes clear: it means *home*, and the warmth of the term is obvious as opposed to the Turkish and Russian word 'yurt,' which simply means a structure or the imprint left by a movable structure.

"Chuulunbaatar was in this valley last year," says Alta, "but no telling where he is now. If his goats are dead from the cold, they won't do us much good anyhow," she says. Alta contacted him and his family a year ago when she was involved in finding herders who might be interested in working on our animal improvement project.

"The *dzud* is killing everything," she says.

"The *dzud*?" I ask.

"Yes," says Alta, "when two or more severe winters are followed by extremely dry, hot summers, like the last three years here. We call it the *dzud*, which also means any calamity concerning livestock. The light snow covers what little vegetation there is, and last summer was a terrible drought—nothing to eat. Last summer, we chose herders to work with us, but how can they work if their animals die? They will need food for their animals."

"If some animals survive," I say, "we will bring them a good buck for breeding. Over time, they will improve their stock."

"When?" she scoffs. "That will take three years. They are dying now." She turns and walks toward the Land Rover.

We continue driving over a range of hills. We stop at the high point of the pass between two valleys. Next to the truck is a mound of stones with a pole in the middle and a blue silk cloth attached. Alta picks up a loose stone from the road, walks to the mound, places her stone on the pile, and then walks around it twice. "You do the same," she says to me as a stern instruction.

"What for?" I ask.

"We must pay homage to those who have passed through here and not returned, and we pray that we shall return this way. See the blue silk tied to a pole sticking out of the rock pile? That is the spirit of those who have come here and died before they could return. Remember them."

I pick up a stone, place it on the pile, and walk around it a few times. Nothing is said, just the remembrance.

We continue driving south, keeping the sun to our right. Many kilometers in the distance, I spot a row of camels. I am thrilled by their timeless beauty and deliberate elegance. "Do those camels mean that a settlement is near?" I ask Alta.

"It may, or they could be many kilometers from their owners. We will follow them. When we reach Chuulunbaatar, do you believe that we are going to convince him to cull his herd when so many animals are dying anyhow?" she asks me. "We should be bringing them a load of hay."

As we drive toward the camel caravan, we see a dead calf on the side of the road, lying on its back with all four legs straight up in the air. It appears to be frozen solid.

Figure 8 Frozen calf blown over by the wind

"When they get weak, they get separated from the herd. Then they stand with their rump to the wind," explains Buyan. "Their legs are skinny, so they freeze first, before the rest of their body.

When the legs freeze, they cannot move, so they stand there until their entire body is frozen. Then, the wind blows them over. Their legs go straight up."

I wonder how many of Chuulunbaatar's animals have died, thinking that Alta may be right that the concept of this project may be meaningless in the face of death.

We keep driving toward the camels, making a massive circle across the desert, slowly turning east and then north, now with the sun to our left. We turn another 90 degrees to face west and finally south again, going 10 km farther south than we went on the northern leg of the circle. We end up farther south than where we started the circle, following the camels. Then we start another huge circle, also trying to follow the path of the camels, making loops as we go. We know where we are only in relation to the last circle and relative to the sun's position. We move south while exploring a circle with a radius of about ten kilometers across as we go. The sun is close to the horizon, but it stays there all day. Even though the sun is barely above the horizon, many hours of light remain.

Our driver brakes suddenly, throwing everyone forward. "Where did they come from?" he shouts.

"Oh, my god!" says Alta in an intense whisper, finally finding something to stir her interest.

Fifty yards in front of us are four men on horseback, all facing our vehicle at the top of a ridge. They appeared without notice, as if from nowhere. The horses stomp their feet, banging into each other with impatience, exhaling steam from their nostrils in the cold air. The riders hold long lances pointed upward with ribbons of various colors waving in the wind behind them. They wear huge fur hats, dark purple or blood-red de'els thickly padded with sleeves about six inches longer than the ends of their fingers to keep their hands warm without gloves, and tall black leather boots up to their knees. The riders have dark complexions, round faces, heavy black eyebrows, and startlingly white teeth.

"What do they want?" I ask.

"They will look for a minute," says Buyan. "Then they may leave, or they might want to talk. It is their territory, so it is up to them."

Buyan steps out of the Land Rover and walks toward them, arm extended, palm up as if making an offer.

"The way Buyan holds his hand up like that," says Alta, sitting in the back seat in the Land Rover, "is a gesture of peace, not hostility. No weapon. He trusts them."

They appear to exchange greetings. Buyan brings his snuff bottle out of a pocket and offers them some. They shake a small amount into their hands and sniff. Their leader, the eldest of the four, reciprocates by offering his snuff bottle to Buyan.

They talk and point while sharing snuff all around.

The elder member of the party points again as if giving directions to Buyan, who returns to the Land Rover, giving a farewell gesture to the horsemen, who reciprocate.

"They are looking for their camels," he says. "I told them we saw many over the last ridge, so they will go look."

"How do they know which camels are theirs?" I ask. There are so many roaming that it seems impossible to keep track of your own.

"Each herder knows how many they have, how old they are, and whether they are male or female, so they only take what is theirs." He pauses for a moment. "I asked them if they knew where Chuulunbaatar was now. They say that a herder family is wintering two valleys to the east, but they don't know if it is Chuulunbaatar. They have not seen him for many weeks," explains Buyan.

He turns to look at the next valley in our direction.

Buyan holds his arm out, pointing his index finger to the southeast.

"He curled his finger like this," says Buyan , curling the tip of his index finger so that it points to the ground, "meaning that they are not far away, less than a day's journey. If his finger had been straight, then they would be far away, more than a day or two," says Buyan.

I glance over Buyan's shoulder out the front window of the Land Rover. The horsemen are gone. I look around the horizon, thinking I can see at least fifty kilometers in every direction. They have disappeared completely.

We alter the direction of our circles to a southeasterly route so that we can continue in the direction the horsemen indicated.

2
The herder Chuulunbaatar's ger: The customs of visiting and eating

On the leeward side of a hill facing the southern winter sun, is the compound of Chuulunbaatar's gers, one for him and his wife, one for his oldest son and wife, and one for three younger children.

Figure 9 Chuulunbaatar's winter settlement

As we arrive, the entire family stands in front of their gers, the children lined up by height. How do they know we are coming? This is my third trip to the Gobi since I started working for the Gobi Initiative. A year ago, I was here at the invitation of the United States Agency for International Development (USAID) with another group that put together a proposal for a civil society contract. At that time, I made four trips to the Gobi to interview herders. Each time, the families knew that we were coming. In the absence of phone lines or cell phones, I never understood how they knew, but they always knew.

A thin stream of smoke ascends from the hole in the center of the ger. Udval, Chuulunbaatar's wife, steps forward very modestly to greet us. I use the name "Udval," though my field notes are ambiguous about her name, and I have not found anyone who can correct it.

"My husband is with the animals," she says. We bow to each other and to each member of the family. "He is preparing a sheep behind the animal pen."

"May we please find him?" inquires Alta who turns to me and quietly says, "It is expected of us to seek him out, so it is preferable to find him before going into the ger."

Chuulunbaatar is behind his herd, kneeling over a sheep. He smiles a greeting without getting up. He lifts the sheep and rolls it onto its back, holding the front legs. He pulls a long, thin knife from his belt, making a short incision in the sheep's underside.

He inserts his hand into the sheep's body, searching inside the animal.

"He is finding the aorta," says Alta. "He will squeeze it, and the sheep will die instantly and painlessly. No blood is lost."

The death is quick, without a bleat. A young man, probably a son, appears to take over.

"Lunch will be ready soon," says Alta. "He wants you to see that it is fresh."

We follow Chuulunbaatar into the main ger and crouch low to get through the door. The pungent smell of the stove's dung fire hits my nostrils. The steam from a large open kettle of boiling water soothes the skin on my face, with a temperature difference of about sixty degrees between the outside and inside. A young woman stirring the kettle gestures toward wooden chairs directly opposite the entrance.

"Those are the head seats for you and Chuulunbaatar," whispers Alta.

I hold my hand as if to say 'ladies first' to Alta, but she quickly refuses. "No, you first. You must sit at the position of honor as the oldest person in our party and the leader of our group," she says as if I am as dumb as the camel dung piled next to the fire. She grabs my arm as I pass and says into my ear, "Remember to pass your snuff bottle."

I offer snuff to Chuulunbaatar, extending my right arm straight out, palm up, with the snuff bottle in my hand. He accepts and reciprocates. We offer it to others in our party. Chuulunbaatar stands up, lifts a small bowl from the table before us, and takes the lid off a large ceramic vase behind us. He scoops a bowl of liquid and turns to face me. He places his finger in the bowl and flicks drops of the liquid to his front, back, and sides, saying something as he performs the ritual.

I lean toward our translator with a quizzical expression on my face.

"He is making an offering to the four directions, north, south, east, and west, and the natural forces of the world," he says. "You don't need to do it, just him."

Chuulunbaatar hands me the bowl, gesturing for me to drink.

The liquid is milky white but almost clear. It is extremely potent with the smell and flavor of rancid milk.

I had a remarkable opportunity to talk with a young woman, Delgerjargal Uvsh, raised in a herder family. She thoroughly enjoyed milking a pregnant mare. "Fermented mare's milk, called airag," whispers the translator.

Figure 10 Young herder girl milking a horse to make fermented mare's milk or airag

Milking a horse is a pleasant chore, the young girl explains. "Among other things, I often milked mares for horse milk that we use to make airag. Milking a mare is easy. They are taller than cows or goats, and it is very pleasant. It is easier to get under a horse. The teats are soft and large; you can milk them every two or three hours. The big problem is that the mare can lose her foal if they don't have enough water at birth, which is scarce in the Gobi.

"My mother and aunt both say it is good to ride the pregnant mare, which is why she always encourages me to hop on, especially when they are pregnant. Light exercise is good for them. A pregnant mare may be sensitive to having a person handle her, and may not be so easy to milk, but I think that is why we were always encouraged as kids to start handling their teats and rubbing underneath their bellies a few weeks before they gave birth. The mare gets nasty if she is not used to feeling something underneath, so she can kick a foal if no one has handled her udders. If you give her affection and attention, she will be less aggressive after giving birth. I think that applies to any animal, including humans. I especially loved being with the animals in the spring. Handling the udders gets them ready for milking to keep them calm.

"It is important for the mare to give rich, nutritious milk full of the mother's antibodies so the foal has all the necessary immunities. Sometimes, the mare can start to leak milk before the foal comes. If she loses too much, all the colostrum, rich with nutrients and antibodies that the foal needs, is gone before the mare gives birth. You can fix that where you have good animal medicine, but not in the Gobi. They lose too many young horses to malnutrition simply because the mare's milk started too soon, so those who want mare's milk to make airag have to wait until the foal is healthy. Everyone is fighting over the mare's milk; before that, everyone is fighting over water. Many animals can go a long time with very little water, but a pregnant horse must have lots of water, as any pregnant animal must, especially just before giving birth."

Alta is thoroughly enjoying this interview as she watches me, transfixed by the young woman's story. It never occurred to me that humans would milk a horse, or a camel, but why not? I am overcome with the reality of being such an intimate part of animal life.

"Airag," says Chuulunbaatar, smiling at his gift to me. I hold the small bowl in both hands, and drink. He scoops another bowl. With nothing to eat for many hours, in sixty seconds, I am holding onto my stool. I have not had anything to drink for weeks, probably not since the flight here. I can feel my head beginning to throb. I am getting hot and dizzy. I focus on the young woman tending the fire. She brings a tea brick, breaks off a handful, and throws it into the boiling pot on the dung fire. Then, she reaches for a bag of white powder and throws a handful into the pot. I concentrate on her movements to steady the room. "Black tea and powdered milk," says Buyan.

Milk tea sounds like salvation to me. After it is served, more water is added to the pot. The young man who took over the sheep slaughter from Chuulunbaatar enters the ger. He holds an armload of animal parts, which he dumps into the pot—fresh mutton.

My head is filled with the smoke of a camel dung fire, steam from the tea, fermented mare's milk, snuff, milk tea, and now, too, the smell of boiling mutton. I glance at Alta, who smiles a knowing grin as if to say, 'OK, *you clever American, now that you know what our life is like, you can start telling us what we need to know from you.*' I am so far out of my own culture that I can hardly grasp my presence, observing myself bearing gifts. But, at this moment, I do not have words to use to tell these people what to do with their lives and their animals. Even having such a thought seems preposterous. I am thousands of miles from any place with which I am familiar, meeting people who spend their lives searching for grass in barren rock and who, many centuries ago, ruled the entire known world. What can I say to them? I feel Alta's curious stare. She knows the words and the ideas because she is part of the team compiling this project. She has been thinking about the animal improvement program for the last two years. But talking amongst ourselves in our office in Ulaanbaatar is a world away from this ger.

The mutton is cooked and served. In front of me are steaming goat organs: the heart, liver, tongue, and brain.

"These are reserved for guests of honor," says Chuulunbaatar.

I admire the parts and decide it is time for more airag. The blessing of fermented mare's milk, which was once a challenge to keep down, has become the nectar of salvation from the sight of boiled sheep organs.

I offer a toast to Chuulunbaatar, one of our three pilot herders, whom the Gobi Initiative interviewed and selected to work with us.

Alta catches my eye with a threatening and hostile glance. "Get out your knife and start eating," she says in a low, intense whisper that is almost a hiss. "No one else can eat, not even Chuulunbaatar, until you start. So start."

A snuff bottle and a large knife are standard equipment in Mongolia. I pull out my pocket knife and cut a sliver of liver as the most digestible treat. This is no country for vegetarians. I become aware that the ger is silent. Everyone is quiet, waiting for me to start eating. "They can't eat until you do," whispers Alta. As I taste boiled goat liver, the ger explodes in motion, cutting, passing, talking, and the joy of eating. What I have prepared to say is subsiding into the pleasure of the feast.

3
The smell of hot brick tea and warm milk

"Alta tells me this is a very difficult winter," I say to Chuulunbaatar, looking at Alta sitting nearby. "How are your animals doing this winter?"

"Twelve sheep and five goats died last week."

"Is there any grass under the snow here?"

"We were told there was grass before we moved here, but there is none."

Alta, listening to us, says, "Last summer, you were much further north when we first interviewed you. Did you come to this place just recently?"

"We just arrived," says Chuulunbaatar. "There was no grass where we were, so we decided to come here. I first asked the other herders if we could move into this area for the winter. The herders said it would be all right," Chuulunbaatar explains.

"Do they give permission?" I ask.

"No, not permission, because anyone can come, but you still ask. It is a sign of respect," he says.

"Are there too many animals here already?" I ask.

"We need the number that we have. My son, Batzorig, his wife Naran, and their child are living with us," he says, looking at the woman near the stove in the center of the ger. "He will need to start his herd soon. I will split my herd and give him half." He pauses, looks away, and says, "I have another son." His face is dark and deeply lined. He has lived in the sun, the cold, and the wind his whole life. His face is stern, but his expression and demeanor are mild. He squints constantly, even when inside, but his look is gentle, as if he sees everything simultaneously. He is not tall. Few Mongolians are tall, but he is broad-shouldered with firm hands. He does not direct affairs inside the ger. He gives no orders and makes no requests. He is a quiet man, thinking over a situation before saying anything but absorbing everything.

"Maybe we could get Batzorig an elite buck, also, if he would like to work with us," says Alta with a sparkle of positivity. "When we were looking for herders last summer, we said that we would bring a very high-quality buck to anyone working with us to improve the quality of their animals. Do you think that Batzorig would be interested?"

Chuulunbaatar stands up, looking at Alta and then at me. It appears unusual for him to stand while inside. Others notice. Udval stops what she is doing at the stove and turns her head to listen. Chuulunbaatar spreads his arms out straight from his sides. The children even stop playing. Everyone is suddenly very quiet. "I agreed to work with you last summer," he says in a slow, direct voice demanding attention, "but I can't use elite bucks if I have no female goats. They are dying. I need hay for the winter, or they will all disappear."

"Can we buy hay?" I ask, turning to Alta, realizing I should know the answer.

"Of course," she snaps, as if to say, *You are the American with money. You can buy whatever you want*. Then she says, "You can buy hay in UB, but the question is, how do we get it to him?"

I turn to Chuulunbaatar. "I will try to find hay when we return to UB, but I don't know how to get it here. I will try."

Alta approaches me and says, "Don't promise anything you can't deliver. Different governments have been doing that to the herders for centuries. The Chinese, then the Russians, and now the Americans. They remember."

Chuulunbaatar is still standing in front of us, not in defiance, nor demanding, nor as a supplicant. He is standing as a man in his own house, having made a statement of importance, a statement of life and death. He stands with more dignity than I have ever encountered in another person, with his family around him, a thousand years behind him, and nowhere to go.

"I know this is a hard winter," I say, "but how have you survived before?"

"The last three years, with bad winters and extremely hot summers, are the worst I can remember. There is light, cold snow in the winter, hiding the little grass here, and extremely hot summers with no rain, so nothing grows. For three years," he says, looking at the floor of the ger. He sits down next to Alta and me.

"If there is not much grass, wouldn't a smaller herd be better?" I ask. "Couldn't you sell the high-quality cashmere fiber for a better price?"

"When the Chinese come to buy cashmere in the spring, they buy it in bulk," says Chuulunbaatar. "They don't look at the fiber. They buy everything."

"But you are not getting the price you should," I say. "When they buy everything, they know they are getting some very valuable commodity in the bulk. I wonder if we could see your animals." I ask. Alta looks at me in surprise. She did not expect this.

"Yes, if you like," says Chuulunbaatar. Alta is on her feet. She would not miss this. She is at the door with her coat on, hat, muffler, and scarf wrapped around her face so that only her eyes are showing, eyes imploring us to hurry now that she is sweltering. She loses patience and goes outside. The rest of us soon follow.

The cold air is a jolt to my airag-addled brain. It sends a shock through my consciousness and clears my sinuses. Then my eyes start searching for light in the black night. The gers glow orange from the stove fires with sparks ascending from the stovepipes. They are beautiful in their low-slung, deep color and vague shape, barely discernible at night.

The glory of the night is the black-on-black moonless sky that reaches one-hundred and eighty degrees from one horizon to the opposite, with more stars than I have ever seen in any other part of the earth. The Mongolian high plateau is at about 5,000 feet in altitude, thousands of miles from the sea, with no dust or interfering city lights to hide the stars. The air is cold and pure, so the stars dazzle my surprised eyes. There are swirls and piles of stars that I have never seen, myriad glowing silver-white dots shimmering in the sky but casting no light themselves. They do not illuminate the ground, yet the sky is full of dark, vibrant, spar-kling light.

I forget to walk. I stand with my head bent straight back, eyes up.

"Alta, I have never seen the sky before."

"It has always been there," she says.

"Can you follow me?" asks Chuulunbaatar. "This way."

A few hundred feet uphill are crude wooden pens for goats and sheep. They are quiet, pushed close together in a clump of furry animals. "The goats are here," says Chuulunbaatar. We straddle the fence and walk to the edge of the herd.

Figure 11 Winter corral

"Do you want a good goat?" asks Chuulunbaatar.

"It doesn't matter," says Alta. "Any will do. I will show you." She walks around the outside edge of the animals, stopping every few steps to pet the spine of a goat and then sliding her hand down the ribcage to the underbelly, gently pulling a handful of hair. "We can go inside now," she says in relief.

I am curious as to what she has. I have heard a description of the coarse long fiber introduced by the Soviets to increase volume, but I am not sure the difference can be seen. We return to the ger.

The difference between high-quality fine fiber and coarse long fiber is easily seen, even without microscopic comparisons. "You need to breed this stuff out," she says with candor and significant lack of tact, holding up some long strands. "If the Soviets did not like Buddhism, instead of espousing the beauty of communism, they burned temples. If they wanted to double agricultural production using their crummy tractors, they didn't improve the tractors. They just built twice as many. If their low-grade steel beams could not hold a load, they forged beams twice as big using the same bad steel. Twice the size does not mean twice the strength, and twice as much does not mean twice as good. If they wanted more cashmere, they bred goats with bigger bucks, not better ones. You now have big Soviet fiber and lousy quality," says Kris, our goat breeder consultant, holding up long fiber strands.

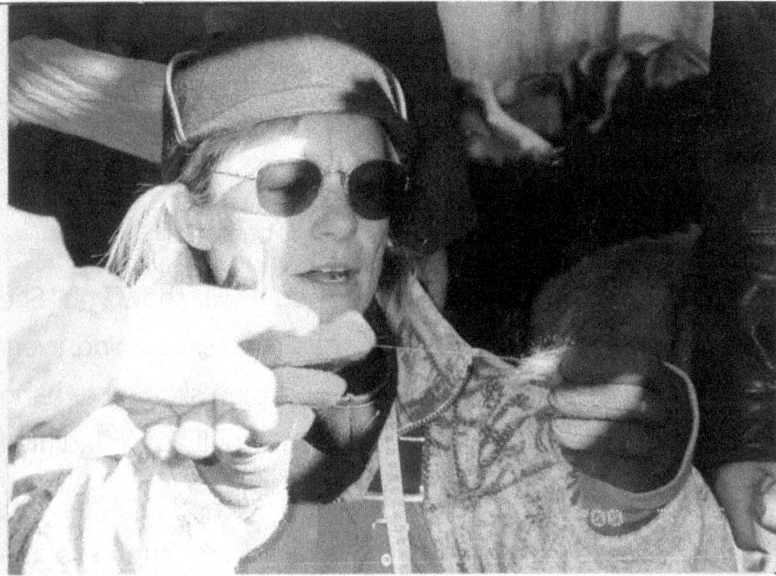

Figure 12 Kris demonstrating long fiber strands.

"We want to give you good bucks to breed out the coarse fiber," I say.

"But it is more than breeding. You are asking me to cull some of my animals," says Chuulunbaatar. "They are what I have, nothing else. A successful man has a large herd. I must take care of my family. A small herd means a man is ready to die."

"The best bucks and the best does will keep your family alive," I say.

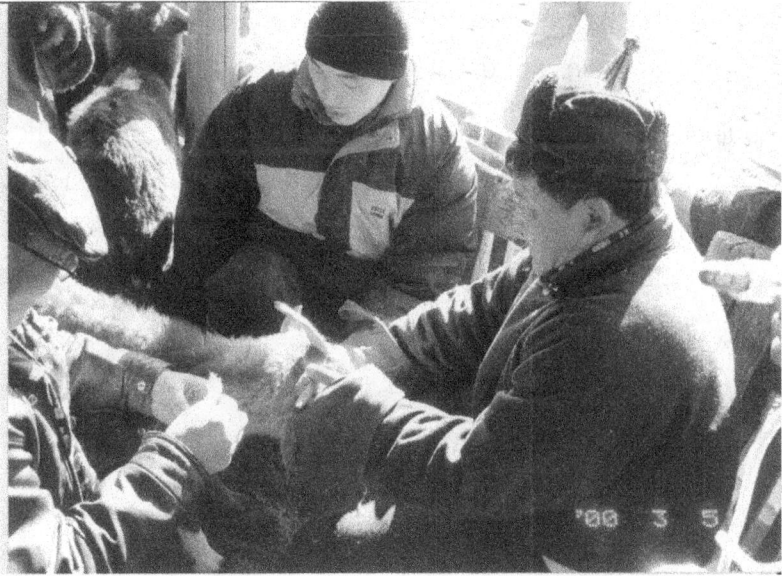

Figure 13 Herders assess the strengths and weaknesses of goat genetics

We stop talking for a few minutes. Chuulunbaatar passes more mutton and airag. Udval replaces the large kettle on the stove with a smaller one. She pours in water and stokes the fire from the dung pile in the middle of the ger. When the water boils, one of the younger women, maybe Chuulunbaatar's daughter-in-law, pulls a handful of powdered milk, pinches off a chunk of black tea, and rubs it between her hands as it disperses into the steaming pot. The smell of warm milk and tea is wonderful. Udval scoops a pitcher of brick tea from the kettle and sends it to everyone, starting with Chuulunbaatar and me, as usual. This time, I start sipping immediately.

Chuulunbaatar gets up again and walks to Udval. They talk for a few minutes. He returns to Alta and me. "Our son is looking for our camels now. He will be back in a few days after you leave. When he returns, he will go to UB and find you. You buy the hay. He will bring the hay to us."

Chuulunbaatar has struck a deal with me. The price of his participation in our animal improvement program is for me to buy hay for the winter. I had better find some. I don't need to look at Alta. She has that expression that says, '*Now you did it. You better deliver.*'

I know what needs to be done. I have promised it, but I am hit with the sensation of terror that I don't know *how* to do it. In the States, sure, but not here. I will figure it out.

4
Chuulunbaatar's wife: Gracious hospitality

I am watching the bustle around the stove, which Udval is tending, while the two young women clean dishes by wiping them with a rag and placing them on a small counter next to the front door. The compactness and organization of the ger are impressive. The women work quickly and efficiently. The mistress of ceremonies is Udval, a meticulous seamstress, sowing the richly colored quilted outer garments called "dels," also frequently spelled "de'el" with the same pronunciation.

Figure 14 My colleague Alta; at the top right, our consultant, Kris; Udval, at the center, with the ancestral shrine behind them in the family ger.

Chuulunbaatar notices me admiring pictures of his family arranged on a dresser top. "My son's wife and my grandson," he says. "My father and mother, and my grandparents," he says, pointing to a group of pictures on the wall, with a few photos surrounded by candles and incense, sitting on top of a low cabinet almost directly behind us.

"My older son is a wrestler, and my younger son is a great horseman, as I was once," says Chuulunbaatar.

"Chuulunbaatar was a national champion rider," says Alta.

Next to the family photos are some faded newspaper clippings with pictures of his sons, either on horseback or with trophies from wrestling tournaments. "My parents," says Chuulunbaatar, pointing to a drawing, not a photograph. "We honor them. We are Buddhist and remember our ancestors. We have been herding here for many generations." He tells me he does not intend to give up or sacrifice his livelihood for any reason. He has obligations to his family and to his ancestors. He will be remembered as they are.

The front door opened with a shaft of light and a huge commotion from two children, laughing and running to Naran, Udval's sister. They are all talking furiously. Naran says something to them that quiets them down. She crouches next to Udval and takes the cooking utensils from her hands. Udval stands, takes a few steps away from the fire, and looks at Chuulunbaatar. I notice her grimace. She remains bent over at the waist as she was while cooking. Chuulunbaatar walks to her, and I follow with Alta next to me.

"Thank you for the food and tea," I say to Udval.

"You were not sure at first, were you?" she inquires, turning her head sideways so she can look at me while still bent over.

I am startled by her frank assessment of my initial squeamishness over the mutton and by her directness.

"How could you tell?" I ask.

"You waited to eat, and then only the liver and tongue, which are always chosen first by non-Mongolians. But the airag helped your appetite, or maybe it softened your resistance," she says, laughing, with a smile. "You will stay tonight."

I look at Alta, my eyes darting around the space, and shrug my shoulders as if asking, "Where?"

"There is always room," says Udval, understanding my question.

"It would be a pleasure if it would not be inconvenient."

She turns her head down, facing the floor.

"May I ask if there is anything we can bring you from UB when we return?"

"Sugar and tea are always appreciated," says Naran.

The front door opens, and the children are full of action again. They run to the door and back to the fire. The door frame is filled with a form, a felt hat, and two huge shoulders that must enter sideways because they would not fit through the door otherwise. After the shoulders, the rest of the body follows, entirely filling the entrance.

A man whom I presume to be Batzorig stands inside the ger in a dark blue dusty de'el and tall hat. He is twice the size of Chuulunbaatar as he faces him. "The camels are here." He turns to me, saying nothing for a moment, looking me directly in the eyes, not intimidating, not aggressive, but with as confident a presence as I have ever felt in any one person before.

"You may see them if you like," he says to me.

I nod to acknowledge his presence as much as to accept his offer. He already knows everything about me.

He turns to his mother, Udval. He must almost kneel to talk to her. His two large hands emerge from the long sleeves of his de'el and reach for his mother's shoulders to steady her. He guides her to cushions next to the wall where she sits. When she is comfortable, he stands and takes a few steps toward Chuulunbaatar and me.

"My father will not tell you because it is a family matter. It is not what you are here to do, but my mother is not well. She went to a nurse at a nearby aimag center a few weeks ago. The nurse found the location of her pain. She has a large lump in her abdomen. I do not mean to impose this on you. My father would not want me to, but our mother is dying."

"Is there a doctor anywhere?" I ask.

"Only in UB, but the trip would be long and very difficult, and a doctor would cost more than we have." He looks at his father, whose face is as still, stern, and cold as the frozen dirt outside.

"When will you be in UB next?" asks the son, looking at me.

"In about three days, maybe four. It is hard to predict."

"If the United States government wants to find hay so our goats won't die, do they also want to find a doctor so that our mother won't die? I will meet you in UB in five days," says Botzorig.

"Can you find our office in UB?" I ask.

"I will be able to find it," he says as he leaves.

We spend the night in what I assume is the children's ger. They must have moved to their parents' ger to make room for us.

5
Morning on the Gobi Desert: How to get out from under piles of felt blankets on a cold morning

The next morning, a thin, white light penetrates the dark ger as a figure opens the door to slip inside. Except for a small opening for my nose and forehead, I am covered under the pile of blankets. The streak of light hits my eyes, which open involuntarily. Our driver Buyan, Alta, and I are spaced around the perimeter of the ger, each of us buried under a mound of felt. I don't recognize anyone. All I can see are lumps. The ger frame is covered with canvas, hides and felt. The materials are of different shapes and sizes, giving the ger a patchwork appearance like an old pair of blue jeans that have been repaired many times. They are all tied down but not sewn together because they will be taken apart, rolled, and packed for the next migration. The wind catches the edges of each piece, giving the entire ger a muffled flapping sound at night as the wind sneaks beneath the obliquely cut pieces. The sound is soothing. With no windows, the ger is very dark.

The person who entered the ger now opens the stove door, places kindling inside, and lights it. The phosphorous smell of the match penetrates the space, and I soon feel the warmth of the fire. The well-bundled person kneeling in front of the stove adds fuel and pours water into a metal pan on the stove. A glove-less hand reaches from inside the long sleeve of the de'el. As the water begins to boil, the person opens a container, pulls out a handful of brick tea, and drops it into the pan with powdered milk. The wonderful aroma fills the ger.

One of the mounds of felt blankets moves with a muffled groan coming from within, like a bear coming out of hibernation. Alta pulls blankets from her face and looks across at me. Our driver's face appears from another pile of felt almost simultaneously. His eyes are dark and wide open, and he is watching the person in the middle of the ger at the stove. Then his eyes catch mine. We look at each other for a very long time, though it is probably only a matter of seconds. His eyes return to the person tending the fire. His face transfixes me. I have certainly seen him many times but have not taken notice. The fire lights his face with a glow that makes his skin a rich orange-brown. I see self-assurance in his focus and peacefulness as he watches the fire. I feel reassured that this man is our driver and guide. I must get to know him better.

He pulls the felt off his body and swings his legs to the floor. How could I have missed this person so completely? He is twice the size of Buyan. He stands and stretches, taking a few steps toward the stove and the person tending it, who notices his presence. They exchange a few words. The figure tending the fire, by tone of voice, is a woman. She stands to face the driver. He nods his head to her. She reciprocates with a bow of her head that is slightly sideways as if to say, it was nothing, and you are welcome. She turns and leaves, letting in a beam of light and a blast of cold air as the space begins to warm from the fire.

The driver turns to look for something under the bedding. He pulls out his boots. They are tall, reaching nearly to his knees, with broad, flat soles, and what appears to be fleece inside, probably lined with lamb's wool. He walks quickly to the door in his long underwear and boots, exiting with little sound. It is a powerful reminder that I have been delaying the necessary trip outside for hours. I bravely remove the warm, heavy covers, pull on boots, and follow him out of the ger.

When we return, Alta sits on the edge of her bed, a felt blanket wrapped around her, grumbling. Buyan is moving, but he is not fully visible yet. Alta has a felt blanket over her head. She stares at the floor, saying something in Mongolian that no one offers to translate for me. Her grumblings follow one more flash of light and cold from the door as she leaves. With this disruption, Buyan starts to move. He pulls the covers from his face and surveys the ger where the driver and I are now sipping tea.

"I wish I knew your name, sir," I say to the driver, who knows no English.

"Naranbaatar, or Naran," says Buyan, springing into his translator role. "His parents are herders, mostly in the Gobi, so he knows the desert well. But he decided not to herd."

"Please tell him I am glad he is with us on this trip."

Naran beams from unexpected praise.

I offer my hand to him, and we shake. I say good morning in broken Mongolian. He seems to understand. His hand is rough and powerful. He drags a canvas bag to a short table next to the stove and dumps the contents onto the tabletop. He unwraps part of an animal's body that appears to be a shoulder and leg of cooked mutton. He gestures for us to join him. He pulls a long, narrow knife like a bayonet from its sheath. He begins cutting pieces of meat from the bone, pushing them toward us with the knife blade. We sit on the floor around the table, milk tea in hand. Naran gives us meat, cutting off the fat, which he sets aside. Alta pushes some of the meat toward him, thinking he is being too polite in giving us the meat.

Naran says something to Buyan, who translates: "He does not eat meat in winter, only fat. That is what his body needs."

He knows how to keep warm, I think. I try some fat. Like fermented horse milk, it tastes appropriate for the place and for survival. Food and drink are becoming part of my life, and I understand the connection between my life and the animals surrounding us. This connection with Midwestern farm families in my background should be obvious. Still, the closeness of grass under the snow, mutton boiling on the stove, and chunks of fat for breakfast give the connection an immediate sense of survival.

We dress in the same outer clothing we have been wearing for days. I admire Naran's boots. "Where do you find those boots," I ask.

"A woman in the aimag center makes them. She makes a living that way. They are horsehide on the outside, sheepskin on the inside."

"No socks?" I ask.

"What are socks," Naran replies. No socks, no gloves, in the coldest climate I can imagine. Socks would only fall and bunch up inside those boots, so why bother? I pull on socks and hiking boots, hats, gloves, and all the other stuff I am now treating skeptically and venture outside.

Chuulunbaatar and his son are waiting for us. Naran starts the Land Rover, which we load with gear.

"My son will be in UB soon, in a few days," says Chuulunbaatar.

"Does he know where our office is?"

"He will find you."

The anxiety of delivering hay, possibly some fodder, and a *doctor* for Udval hits me. "Alta, we have work to do."

"Yes, you do," she corrects me. It is my promise to fulfill, not hers, though I am convinced she hopes for success. These expenses are not in our budget, but I am thinking that if Chuulunbaatar does not make it through the winter, the animal improvement program would be severely reduced and our project compromised. A case will have to be made to our Chief of Staff and to the USAID director in UB. These are office people, not field people. They connect program goals with benchmark performance measures over a five-year work program. I am playing out the conversation in my head already. We also do not have money for health care for the herders' family members. The survival circumstances faced by our pilot herders do not match well with our five-year animal improvement plan. I repeat that thought and become conscious of a horrible coincidence. Decades of fruitless Soviet planning always revolved around five-year plans, which never caught up with the doom of the crisis at the time. How many of those plans were promulgated during the Soviet era, endless five-year plans that never dealt with human tragedy confronting peasants, workers, or herders daily? What am I doing that is different? I have a better plan, but it is still a five-year plan, so in the meantime, what will happen to Chuulunbaatar's wife and animals?

"I will meet your son in UB," I say to Chuulunbaatar. "Is your wife inside?"

"Yes, but she does not feel well."

"I will do what I can to find her a doctor. We can talk about it in UB." This is another promise that I must fulfill without having any idea how. I don't even look for what I am sure is astonishment on Alta's face.

The truck is warm and humming. We say goodbye. I ask Chuulunbaatar, "Who had started the fire and fixed the tea for us"? "Ariunaa, my son's wife," he says, turning to her. Alta whispers in my ear that the name means "pure" and is a very auspicious Mongolian name.

I look for her among the family members congregating to see us off. Finding her, I say, "Thank you especially. This morning's fire was appreciated, and the tea was the best." She smiles and acknowledges.

Walking to the truck, I say, in a loud voice, "Shotgun." Alta reaches for the front door on the passenger's side. "I said shotgun," I tell her. "I get the front seat next to the driver."

"Why? What is shotgun?" she exclaims, annoyed.

"You know, in Western movies. Have you seen a stagecoach?"

"Of course," she says.

"Well, the guy next to the driver always holds a shotgun. That is his job. So, if you want to ride in the seat next to the driver, you have to yell shotgun first."

"You're crazy," says Alta.

"That is the American custom," I say, laughing and entering the front seat.

"Where is your shotgun?" she asks, grumbling as she climbs into the back seat.

"It is only a figure of speech."

"What is a figure of speech?" she asks, even more annoyed.

"I just taught you one."

6

An unobstructed view: Travel in the Gobi and order a pair of custom-made boots

Figure 15 Wild horses on the Gobi Desert

Finding UB will be easier than finding Chuulunbaatar. UB is where we left it. It doesn't move around. We are looking for the road north that will take us there. I am sitting in the back seat, having lost the shotgun argument to Alta. "You make up the rules to suit yourself," she says while climbing into the front seat, "and not just with a shotgun," she mutters, not looking at Buyan or me in the back seat.

"I don't have rules, just goals," I retort as we start the bouncing, vibrating, frigid return to UB. The wind pushes the Land Rover sideways as the wheels spin on the frozen sand, which offers no traction. "I just want the herders to have healthy animals, to breed their animals so that the fiber will get better over time, to sort the fine fiber from the coarse, and to market the good stuff with other herders so they all get a fair price. They need to learn to work together, to cooperate," I lecture.

"Before they do that," Alta says, "they have to take care of Chuulunbaatar's wife Udval and feed their herd. But I bet you have a rule that you can't use your money for hay and doctors. Everyone makes promises. Do you think that the U.S. government cares about cashmere goats? What do the Americans want here?" she asks.

"This is a hostile world. Americans want friends. If that means taking care of goats, we take care of goats. If our motives are wrong, why are you working for us?" I ask.

"Because I want the herders to have healthy animals, too, so if that is why you are here, I will work with you."

"Alta, what I want and what I can deliver to the herders might differ. Whatever money we have is controlled by Ed Burgells, chief of the USAID office in the U.S. Embassy, not by me."

"I know," she says, "but you make promises as if you have money. ***People remember unkept promises longer than they remember gifts.***"

She is right. I need to acknowledge my limitations. I will likely do something and use that statement to argue for getting the funds I need to fulfill the promise. Alta is telling me that that is a risky approach. What if my arguments fail? The herders are hurt, not me.

After three hours of wandering and challenging driving, we start seeing gers in clusters, some fenced areas, and abandoned cement block buildings that are remnants of the Soviet period. Slowly, a town center appears. The cement buildings in the town center look the same as the abandoned ones on the outskirts of town, except those in the center are occupied. We stop on a slab of concrete across from a commercial building. In the middle of the slab is a lone gas pump. We stop next to the pump. Our driver tries the pump but has yet to be successful. We all leave the Land Rover as Naran pulls a toolbox from the trunk. He spreads out some tools on the ground and starts to tinker with the crank on the pump, which does not move. He points to some tools, asking me to hand them to him. As I pick them up with my gloveless hands, they stick to my moist skin, which was buried deep in my pockets. My hands ache from the cold tools.

Naran's efforts are fruitless.

A shopkeeper across the street tells us the owner will be here sometime this morning with tools to fix the frozen gas pump. Gasoline has water in it that freezes, just like a well pump. The wind rattles the sheet metal on the gas pump and rocks our truck, whistling into any window or door crack. It feels well below zero.

We notice an outdoor market a few blocks away, so we tighten our coats, angle our bodies into the wind, and head down the road. The market is an intoxicating and thrilling hit for me. It is mostly outdoors but protected by sheds, lean-tos, and canvas strung on poles. Charcoal fires offer roasted goat and boiling pots of tea and mutton, sending steam everywhere in the wind. The smells and steam, the people in brightly colored quilted de'els with large fur hats all mix with tables of pots and pans, clothes, soap, tools, utensils, and kerosene to form a joyous sense of warmth and place. I start looking for boots like the ones the herders wear, tall boots almost up to the knees with flat soles and toes pointed upward. I ask a woman who is selling clothes and shoes about boots. She looks at my feet and laughs, calling her friends to see them. Four or five other women gather around her to look at my feet. They all start laughing and pointing.

"Buyan!" I exclaim, or more like a cry. "I need help."

"They have never seen such big feet," he explains.

Buyan talks with the women until all of them, including Buyan, are laughing hysterically. The group is about ten or twelve, with a few kids crawling on the ground to see what is so funny. "They say you must buy a horse, slaughter it, and tan the hide to get enough leather for your boots." As he translates what he just said to me into Mongolian, what is now a crowd starts howling with laughter.

"Buyan!!"

"OK, OK. We will go to a bootmaker who lives here in town." He talks with the women who started the festivities. "The boot-maker learned her trade from her father, who made boots until the Soviets prevented him from working so everyone would buy surplus army boots."

The bootmaker is a lovely, shy young woman living in a house with many other people, including her brother. She cares for a small child but does not have a husband. Buyan says that she very much needs the income from making boots. "She seems to be supporting herself. She would like to make you a good pair, the best you can find," he says after a brief conversation.

We follow her to a workroom piled high with hides, sheepskin, rubber, thread, and tools on a table. She gestures for me to sit in the only chair. "I will make these boots just for you. What would you like?"

"High-topped boots like the herders wear."

"Here are some kinds of leather. What would you like? Horsehide is the toughest, and I have a nice piece of hide that will wear well. Then felt and rubber soles, all lined with lamb's wool?"

"Yes, excellent."

"And on the side of the boot, a symbol of Mongolia?"

"Yes, the figure that I see in some places, like the Norwegian knot," I explain. We find a picture of the symbol. The bootmaker likes the idea and says she can make it in leather.

"It will take me two weeks. Would you deposit the materials?"

"Of course," I tell her, giving her more than enough for the deposit. "How will you get them to me?" I ask.

"I will give them to someone from this town who is going to UB to deliver pelts. You can give them the rest of the payment."

She lays a piece of paper on the floor and asks me to step on it. My foot doesn't fit on the paper, so she places another piece on the floor. "You have a large foot," she says with some dismay. "I never made such a big boot," she says to Buyan in Mongolian, implying that she does not want the comment to be translated. Buyan will tell me later.

"I hope to make the best boots in the Gobi Desert. You will like them," she says. Her household of six or seven people and a few children is jammed in the doorway to her studio, watching.

After she is satisfied with the measurements, Buyan and I walk to the Land Rover, waiting by the gas pump. Unfortunately, as we approach, we see Naran and the pump man kneeling on the ground with parts and tools. The wind is stronger and colder. The idea of handling metal parts in this cold seems impossible. "Where is Alta?" I inquire.

Buyan points to the Land Rover. She is in the front seat under many coats with a hood over her head. I can barely see her nose—no need to ask her how she is doing. Neither the engine nor the heater is on. I see her breath coming out from a small hole in the coat pile on top of her. I leave her to her misery.

An hour and a half later, the engine is humming again; the main tank and all the spare tanks strapped to the rear bumper are full, the heater is blasting, and all four wheels are spinning on frozen gravel again. We head north through three primary valleys of the Gobi Desert toward UB. At the first high pass, we stop to place stones on the road marker, as we had done before. The stone piles are very high, with a pole and pieces of blue cloth strung from them. Many travelers have passed, possibly thousands of travelers, over thousands of years in a country where everyone moves constantly. The simple monument expresses hope that you will remain safe to pass here again. Walking around the stone pile, I ask Buyan if he says a particular prayer or words.

"No. It is more a Buddhist meditation, the highest form of prayer. Meditation is an internal seeking for safety or wholeness, not a request to someone else. I can make myself safe. No one else can do that."

I look at the valley and the unending horizon that could be a thousand miles away. The sky and sand meet somewhere, though I can't tell the difference. I see no fences, signs, or boundaries. We are only aware of direction based on the sun's travel across the sky. Especially for a nomad, travel is a significant risk, explains Buyan. "A safe return this route is what we meditate about, not the destination," says Buyan. "Movement is life, not staying still." The view is complete, unimpeded, unobstructed. I place another stone on the pile, hoping we all return this way someday.

Figure 16 Forever view of the horizon and a camel train

7
A bowl of soup: Classic Mongolian roadhouse

A few miles outside of UB, the trail becomes a paved road with a few outposts, fuel storage tanks, and an occasional group of buildings that appear to cater to travelers who park their trucks in front of the buildings. Five or six cement block structures line the road. They have no menu, but Naran seems to know what he wants. He drives slowly in front of the buildings on one side of the road, then crosses the road and backtracks behind the buildings on the other side. He finally stops in front of one building. They all look the same, but this is the place he chooses.

We walk inside through a short, narrow doorway. The sun outside is piercingly bright, a white-blue color like cheap fluorescent lights that are cold, not warm.

The inside is so dark that I cannot see anything at first. It is just as cold, but for some reason, the dark makes it feel warmer than outside. Just being out of the wind must make the difference. A stooped-back older woman looks at us through a half-open door, saying something to Naran that sounds like recognition or greeting. Then she counts us. We sit at the three tables that fill the room. She disappears. Ordering is not necessary.

"Mutton soup," says Alta. With a nod, the cook subtly declares that she has enough soup for all of us. A young boy quickly appears to place charcoal on the stove in a corner of the room. He lights a fire, and the space begins to warm.

We sit quietly. After many minutes, the cook brings us giant bowls of steaming soup. The steam washes my face, the smell of boiled mutton, potatoes, maybe some onions, carrots, and a grain, possibly barley or another hardy grain that grows in the arid north, forming a luscious blend of smells: charcoal, mutton, vegetables, steam, warm wool, and close humans. I realize the vegetables are all from underground, like carrots, turnips, or potatoes, so the ferocious summer heat does not scorch them. I inhale the steam and smells, just wanting to absorb the whole of it. I wrap my arms around the bowl, hugging its warmth and holding it within inches of my face. I keep the bowl with both hands, slurping appropriately to show appreciation. The soup has tiny pieces of mutton, mostly gristle or fat, which is now most appealing. The cook places a plate of hard bread on each table. We dip the bread into the soup to soften and devour it quickly. She brings a kettle to each table, and she offers refills.

Her expression changes from when we first arrived when she looks worried or anxious. Her face is now at ease, relaxed. Her eyes were squinting, not wide open, as when we arrived, she pushed her lips out, puckered from her teeth, and not curled into her mouth. The room is warm from our presence, the soup, and the fireplace in the corner. She looks satisfied that she can serve us. She stands in the doorway to the kitchen, watching us and staring at me. How many Americans have been in her place? Very few, if any, I suspect. It is a familiar experience from China where staring is done with a penetrating and uninterruptible gawking for many minutes, not as a contest as to who will blink first, but just out of the sheer pleasure of curiosity. It is not impolite. It is the same in Mongolia. She is engrossed and will only quit once fully satisfied that she has absorbed every detail. I can feel the questions in her eyes: who is this white man, and what is he doing with Mongolians in my roadside cafe? I cannot explain why I am here. She drops her eyes as she becomes aware I am also looking at her.

"We are working with the herders," says Alta to the woman, "to make the goats stronger with better fiber for the market."

The woman looks at Alta and smiles. "Alta, you explained that beautifully," I say.

"Of course," she snaps, "We understand what you are doing." Alta then says something that she should translate but does not. The woman laughs, turns, and retreats to the kitchen.

"Alta?" I ask, annoyed.

"Time to go," says Alta, looking at the others, ignoring my question. Naran pulls money from his pocket and sticks his head into the kitchen, saying something to the woman, followed by a salutation. The woman comes out of the kitchen to bid us farewell. I can tell from her expression that we have paid her well.

On the way to the Land Rover, Alta turns to me and says, "She comes from a herder family. Everyone here comes from a herder family. Everyone understands that there are too many animals on the land, but no one will change on their own. It is the difference between knowing and doing."

I know what I have to do when we get to the office. I must explain the need for hay and a doctor to Stephen Vance and the USAID Field Director." Vance is our Chief of the Party, and the USAID Field Office is the agency of the U.S. State Department that funds our program. They must approve any unusual expenditure, which may be challenging to accomplish.

"Can we find hay in UB?" I ask Alta.

"Of course, but if we don't, you are in trouble."

"I am in trouble either way, with Vance if we have to spend money that is not budgeted and with Chuulunbaatar if we don't."

"You will figure it out, boss," she says, with a distinct tone of pleasure. She relishes using the word 'boss' when it means doing something she is glad not to do. It means, OK, big shot, it is up to you.

Just as we reach the truck, she stops, turns to face me, takes a deep breath, and at the top of her lungs, she yells, "Shotgun!"

8
The late Mr. Stephen Vance, Chief of Party: Taciturn brilliance

Returning to UB is a luxury after traveling in the Gobi. My austere concrete apartment building is now a spa. I skip up nine floors with ease. My rooms are warm, bright, and cozy with multicolored rugs on the floors, on-demand hot water, a toilet that works, and a small yet helpful kitchen with cases of bottled water, a stove, a refrigerator, and a table, and chairs with a view of other beautiful concrete apartment buildings. A shortwave radio on the kitchen table connects me to the world, but no newspapers or TV exist. I search around the dial for BBC World Service. Saturday morning is 'Steven Wright-round the World' with a light-hearted and eclectic international call-in show. This place is my little piece of heaven.

On the dresser in my bedroom are many pictures of Laurie and the small metal holder and candle for the "day of confidence," our day of intimacy and trust in the future. It is our special day, the twelfth of the month, for which we each light a candle. We had met a month before Mercy Corps offered this position to me. I know that Laurie is the person I want to be with, but I applied for this contract with another group more than a year ago. Our application was one of three finalists, so the United States Agency for International Development gave us fifteen thousand dollars to come here and perfect our application.

I had to tell her I would soon be leaving for Mongolia. We agreed to light a candle on the twelfth day of each month, marking the day of our first date. My candle waits on my dresser in UB to be lit on the next twelfth day.

The next day, even my office looks inviting. I move stuff around on my desk to touch it. I write something in an email to Laurie, then dig out my notebook and begin to draft a field report for Vance. After any trip, our Director requires us to file trip reports, which are reviewed internally and then sent to Ed Burgells at the USAID office so he knows what is happening. My report is full of urgent funding requests.

"How can I send this trip report to the Field Office?" says Vance as he walks into our office holding up my account of the trip to Chuulunbaatar's ger. "The money for hay is not in our budget. We can't buy hay for every herder in Mongolia. That is not our job," he says as he flings the report onto my desk.

Vance looks at me with disdain and says, "You expect me to ask USAID for an amendment to our program contract to provide medical service to a herder's wife, to purchase hay, to speed up the purchase of elite bucks as part of the breeding program, and you also want to bring as many herders as possible to UB to introduce them to western marketing techniques.?" he says with disdain.

I feel challenged by my supervisor, but I have an innate resistance to authority when he contradicts me on a matter that I consider essential to the success of my mission. This interchange is a critical moment in the power struggle within our organization. I believe fervently in the value of cooperation, which is why I am here to work for Mercy Corps. But I also know what we need as a team to work successfully with our herder colleagues. We are asking them to make sacrifices, and for them to do what we ask, they are telling us what they need. And my word is at stake. But I am not asking for personal reasons. What may sound pompous, but I represent the United States of America, and we should do everything we can to meet the needs of our herder partners.

I will also admit that my initial impression of Mr. Vance is that he is an accomplished, even brilliant, dedicated public servant, but in the context of being an efficient bureaucrat. I fear he is more concerned about keeping the Gobi Initiative on budget than making dramatic social and economic changes in this spectacular country. As we will later discuss, first impressions are often misguided, and a set of tragic circumstances will alter my view of my supervisor.

However, returning to the story, I respond.

"If you want to have an animal improvement program this summer, yes," I say. I am not going to be apologetic. My three colleagues in the office are quiet. I glance at Alta, not wanting to catch her eye but knowing that her gaze pierces me. Our eyes lock for a moment: "See, see!" Her eyes say to me, "I told you so."

"I think we need to discuss how we can accomplish this," I say.

"Yes, let's," Vance responds. "In an hour, my office." As Vance leaves, Alta walks toward my desk.

"You don't need to say anything," I tell her.

"I am not saying anything," she says. "Yet."

"Why would I say anything? He told you what you need to know. Has he ever been out of his glassed-in office? Has he ever visited a herder's ger in the desert? Maybe we could take a video camera next time to show him moving pictures so he…"

"Alta!"

I hear a ring on my computer, indicating a message. I open it to an additional instruction from Vance. "I just spoke briefly with Ed," reads the email message. Ed is Vance's boss as the USAID manager at the U.S. Embassy in Mongolia. "Ed says that the only reason Chuulunbaatar wants hay is so that he can sell it for a profit. You have to prove that there is no hay available and sign a contract with him stating that he will not sell anything that we give him. Write me a proposal on accomplishing these things before EOB today," signed Vance. "EOB" is Vance's favorite bureau-cratic abbreviation, much better than ASAP, which could be anytime, depending on how you define what is possible. EOB means by the End of Business for that day and is a precise time-specific order.

"Will do," I respond. How can I prove that there is no hay in Dundgov's aimag? How can you prove that anything does not exist? You can assume that nothing exists, so when you find it, you refute a null hypothesis but you cannot prove a null hypothesis. Vance's request is impossible to fulfill, but maybe intentionally so.

On the other hand, the idea of a contract is promising. That makes sense. Shall I send Alta into the Gobi Desert to look for hay, and when she finds none, we can go to the Embassy with our funding request? How long should I ask her to look? Can I send her there for a month?

"Alta," I yell across our office space, "I have an assignment for you." It is the end of the day after business, and she is gone. We will start this project fresh tomorrow.

9
International Politics

Altantsetseg Banzragch (Alta), Program Officer for Agriculture, is my trusted colleague and Mongolian counterpart. All American program managers have Mongolian equivalents, so the operation of an American company that has relocated to Ulaanbaatar, Mongolia, looks different. She is sitting at her desk as I walk into our office. She looks up as I approach.

"So, Alta, where are the dead bodies today?"

"What?" she says, startled, looking up at me.

"It's just an expression."

"Crazy American. Another figure of speech?" she mumbles, returning to her work.

"We have to find the villains, the bad guys today," I say.

"Of course," she says. "I will start looking right away. (Pause). What are you talking about?"

"Vance and I just discussed our request for hay, doctors, elite bucks, and medicine for the herd. He says that Ed Burgells thinks that all the herders are thieves, that all they want is money from Americans, and that they will take the hay and sell it to their neighbors for a nice profit. We have to prove that there is no hay for them."

"I'll start looking this afternoon, but the Gobi is large. It will take about five years," she says.

"Maybe it would be more efficient if we ask around," I suggest.

Alta stands up at her desk. "Ariuna!" she calls to our colleague working at another desk in our office area. "Have you seen any hay today?"

Ariuna looks at Alta with the expression she always has for Alta as if to say, "Now, what dumb thing are you going to say to me that is going to get me in trouble and lose my job? I would just as soon not be associated with you, but I will tolerate whatever stupid question you have."

"No, I have not seen any hay today." Ariuna is a tall, slender woman in her mid-40s with a soft face and a countenance of tranquility and humor. She is sincere and lighthearted, whereas Alta is realistic and skeptical. Ariuna always has the hint of a smile in the curl of her lips and sparkling eyes. She hasn't a drop of sarcasm in her entire being. She has excellent English and a good sense of organization and is responsible for travel arrangements and logistics. She knows where we are and what we are doing at all times but does not care why we are doing it.

"There you go," says Alta, turning toward me. "No hay here."

"Parliament. My office. Five minutes," orders Vance, who is standing at the entrance of our office area, very possibly observing our investigative repartee.

"Alta," I ask, "while I talk with Vance, see if other Gobi Initiative staff have ideas about how to find hay in the Gobi Desert or confirm that there is none available."

Knowing this request is critical for our efforts to convince the USAID chief of the need for funds, Alta turns to her desk, looking like a woman on a mission.

Five minutes later: "The Ambassador would like to discuss your last trip with you and me at the Embassy this afternoon. Can we leave by 2:00 p.m.?"

"We certainly can," I say.

"And I have an idea about searching for hay. Let's talk to Layton to see if his contacts in our aimag in the Gobi can get the word out that we are looking for hay. Through the newspaper, we can see if anything is available," Vance suggests.

"Excellent," I say. Layton Croft is the editor of the Gobi Business News, published by our agency and distributed throughout all of the aimags in Gobi.

Alphonse F. La Porte, U.S. Ambassador to Mongolia, is stashed away in a concrete bunker of a building, an excellent example of 1950s Stalinesque architecture. It is a two-story box surrounded by a chain link fence on top of a three-foot cement block wall. Outside the fence are two rows of concrete road dividers like the ones used during highway repairs to keep traffic separated. After the October 23, 1983, car bombing of a Marine Corps barracks in Beirut, Lebanon, the U.S. keeps motorized vehicles further away from U.S. installations in foreign countries.

"We are here to see the Ambassador," says Vance in a bureaucratic tone that sounds practiced but has become part of his persona. "We have an appointment."

Vance and I produce our passports and Gobi Initiative identification cards. The U.S. security police are skeptical, but we phone the Embassy to confirm our claim. The response on the other end must have been curt and direct. The security guard's backbone stiffens. He sits up, snaps a look at us, and almost salutes the phone. "Yes, sir, immediately".

Turning to us, he says, "I will keep your identification documents safe while you are here. Please retrieve them on your way out. Here is a receipt. Please fasten this pass to your lapel and keep it in plain view during your visit." He gestures toward the door leading to the Embassy entrance.

"Hello, Vance," says the Ambassador, walking briskly toward us as he welcomes us into his office. He extends his arm for a vigorous handshake. "And this must be Mr. Parliament."

"The same. Good to meet you, sir."

"You had quite a trip to the countryside for your first official visit. You were in Omnogovi or Dundgovi? Remind me." Omnogovi is Mongolian for South Gobi Province, which is on the border with China, and Dundgovi is Middle Gobi Province, the next aimag to the north.

"Dundgovi for most of the trip. We went through Mandalgovi, the aimag center. But our pilot herder, Chuulunbaatar, had moved far to the southeast. We may have been in Omnogovi or

Dorngovi when we found him. The exact locations are challenging to judge".

"I heard he has already lost nearly 500 goats from the dzud," the Ambassador says. "We must keep his good goats alive so the breeding program can work."

This Ambassador knows what is going on out there.

"We are concerned about creating something of a windfall profit for him," adds Vance. "We don't want to send hay to someone who does not need it or might even sell it for a profit."

"Is it clear to us that the herder needs the hay to survive?" the Ambassador asks, looking at Vance and me. "Isn't this where we have to trust someone?"

"I believe so," I say. "The herders' lives are much more difficult than I had imagined. And Chuulunbaatar's wife is very ill. He doesn't know from what."

"Well, get her in here. We will check. How much do you need, Vance?" asks the Ambassador.

"We don't have an estimate yet. We can probably find hay in UB, but the cost is in getting it delivered. We will give you an estimate," says Vance, looking at me as I nod in concurrence.

"Now, let's talk about that border with China," says the Ambassador, offering us a chair, as we have been standing for this conversation, which to me was critical, but to Mr. La Porte was inconsequential and also, apparently, not the primary matter on his agenda. "Right now," asks the Ambassador, "who is the single purchaser of raw cashmere from the herders?"

"The Chinese," I respond.

"Exactly, at any price they want to set."

"Single buyer, no competition," offers Vance.

"And they pay no tariffs to the Mongolians because the border is so long and impossible to police," continues the Ambassador. "A year's salary for a border guard is nothing to a Chinese cashmere smuggler. Instead of cheap raw cashmere fiber going to China, we want to see it come to UB, to the Mongolian processors. What do you think of a united front against smuggling into China? Could we get together Mongolian trade officials, police, the military, the foreign affairs office, and some representatives of the herders to encourage the Mongolian government formally to request China to stop illegal cashmere imports into China?"

"What do you think of asking the herder associations to represent the herders to the government?" Vance asks, looking at me.

"Our best contact is probably Nadmid at the Association of Mongolian Agricultural Cooperatives and Tsendmaa at the wool and cashmere federation. However, those cooperative associations are left over from the Soviet era and may not represent anyone now. Tsendmaa is more of a promoter of Mongolian products than a representative of herders. Maybe we should bring our pilot herders directly to UB to meet with government officials face to face," I say.

"We had a meeting like that in our original plans when we started the Gobi Initiative," says Vance.

"Yes, a direct meeting, face to face, with Mongolian officials and herders. You see, there is another complication. We know that the

Chinese want cheap raw cashmere, but the U.S. cannot approach the Chinese to stop smuggling. According to international trade protocol, the country being harmed, in this case, Mongolia must initiate pressure on the offending country. It is not a U.S. issue. The pressure on China must come from Mongolia."

"Why would the Chinese cooperate?" I ask.

"The timing could not be better," says the Ambassador. "A resolution is being prepared for the U.S. Congress to offer most-favored-nation status to China, which is the critical step in their application for membership in the World Trade Organization. For years, they have wanted to become a member of the WTO. Still, for the resolution to pass, China must respect all international borders, and they cannot subsidize their own processing companies because the WTO has a policy against subsidizing exports. It is an unfair advantage."

"How close are they to being admitted?" I ask.

"I understand that the negotiators in Switzerland will conclude their discussions within a year. Then UB processors will not have to compete with the Chinese government on the price of raw cashmere, so we should see more products staying in Mongolia."

"I think we can explain this to Chin-Erdene at the Mongolian Foreign Investment and Trade Agency and Davaasambuu in External Relations," says Vance.

"We don't want to close the border; we just want fair trade all around," says Ambassador La Porte. "Why don't you see if your friends at the Japan International Cooperation Agency would agree?" asks the Ambassador, referring to the Japanese development agency, roughly equivalent to USAID. "Let me know how it goes, and I will discuss the matter of purchasing hay with Ed," he says in closing, clarifying that our time has expired.

The link between the Ambassador's foreign policy objective of using Mongolia to push China on trade policy and finding a few dollars for hay was unmistakable. It is a reasonable tradeoff.

"So, Vance," I ask as we drive to our office, "how are our relations with the Japanese?"

"I have met them, but only briefly. They have a mission here in UB called JICA. I think the Ambassador referred to it."

"Yes, I was not sure what that stands for."

"Japan International Cooperation Agency, just like our USAID," he says. "They are very low-key but influential. They are investing in Mongolian cashmere processing plants here in UB and trying to purchase raw cashmere fiber that they clean and spin in Japan."

Returning to our office, I ask Ariuna to make an appointment with JICA. "I will make an appointment for you, but I am not sure where they are," she says.

In her typical efficiency, she finds them, makes an appointment for me, and gives directions to our staff driver, who takes me to their office. "You would never find it on your own," she says, smiling.

"I wonder if I should bring a translator. Does anyone know Japanese in our office?" I ask.

"No, no one speaks Japanese here. You will have to rely on their translator if they do not use English," says Ariuna. "Maybe we had better find someone for you. You never know what the translator is saying," she says, obviously concerned.

"No need. Our meeting with the Japanese is to get to know each other, and they will have a translator in their office," I say in a hopeful voice.

"The first meeting can be the most important," says Ariuna. "It lays the foundation."

I am always confident in my social skills, projecting sincerity and assuring others that I am not a competitor but a person looking for solutions. As I think about it, however, high-end textile and clothing manufacturers in the United States are interested in purchasing raw Mongolian cashmere, the same commodity the Japanese seek, so we are competitors. If I think about it from the Mongolian point of view, they want to get the best price possible on the international market. That would be good for the herders and the processors in UB, consistent with our Gobi Initiative mission to stimulate indigenous economic development. So, whose interest am I pursuing? I must stick to the Gobi Initiative mission, which is not in the interests of the Japanese or the American importers but ultimately in the Mongolian herders' interests. I am not so sure that the State Department views it that way.

"Vance," I say with a false bounce as I check into my supervisor's office before venturing to meet the Japanese, "what is our experience with JICA? Should we work with them in a partnership? I noticed in my work program that the staff wrote before I arrived that we were supposed to hold a cashmere summit."

"Let's be careful," he says after some thought. "But we should retain control of the planning."

"Well, I think we are prepared to do that," I say.

"Yes, then we can set the format, but we should also keep the Mongolians here in U.B. in charge, in the public view. We do not want this to look American-run," says Vance. "I have not met him, but I understand that Mr. Hara is the key person at JICA. He is influential, and he is a good negotiator. JICA is behind everything here. Japanese businesses are investing heavily in Mongolian processing companies. They want cheap fiber."

"Same as the Chinese and Americans," I say.

"Get what you can, and don't make any promises," says Vance.

"Do you want to join me?" I ask out of deference.

"I think it's best if you represent us for this meeting. We don't want to place too much importance on it yet," Vance says, not acknowledging the slight to my inferior status.

In a quick moment of reflection, he is correct. We send staff people first to work out details. The director can come later, either to sign an agreement or to make a formal proposal, if it comes to that. Our mission is clear. The Gobi Initiative is on the side of the herders and local processors who take the coarse cashmere fiber, wash, card, and sometimes spin it into thread before the processors export it. The Japanese wish to buy thread or washed and bundled cashmere fiber, looking something like bales of hay. The latter is cheaper per pound because it is not significantly value-added by the processing cost of making thread. However, it is more advantageous for the Mongolians to make the thread because it is much cheaper to ship thread to Japan than bulk fiber. Whichever is best for the Mongolians to export: raw unsorted fiber straight from the herders, sorted, washed, and carded bulk fiber, or processed thread, is up to the Mongolians.

With that briefing, I am off to meet with the Japanese economic and trade delegation. The JICA delegation has a low profile in UB, but it is three times the size of the USAID staff.

"Mr. Parliament, allow me to introduce Satoshi Matoba, our Second Secretary for Economic Cooperation; Hiroshi Fujimoto, our Third Secretary from the Japanese Embassy; Kiyoshi Hara, a consultant to JICA on small business development from the Ministry of Agriculture and Industry in Tokyo; and Shinzo Tanaka, who is with the Trade and Development Bank of Mongolia as provided by JICA," says the polite and gracious hostess and translator.

"It is a pleasure," I say as I shake hands with everyone. The assemblage is not a delegation; it is a diplomatic army. I am outnumbered. Mr. Hara, a U.S.-trained economist, becomes the spokesperson because of his expertise and excellent English.

"The goal of JICA is the same as that of your government, I believe if I understand the USAID mission here," he opens. "We wish to help develop the Mongolians' cashmere enterprise."

"It would be best if all the cashmere from the Gobi came to U.B. for processing," says the Third Secretary, Fujimoto.

"Because Japan has a large garment manufacturing capacity, they can purchase all the processed cashmere that Mongolia can produce at a fair market price," says Mr. Tanaka, "but now none of it is coming to UB. It is going south."

He means China, which has an even larger garment manufacturing capacity and is, it occurs to me, a direct competitor to the Japanese in value-added products.

"We are talking to the processors in U.B. about getting the Mongolian government to close the border for some time until the local industry is up and operating better," continues Mr. Hara.

Enforcing border tariffs and getting the Chinese to stop allowing smuggling would accomplish the same thing.

"Impossible," retorts Fujimoto, who, being from the Embassy, is on the foreign policy side of the discussion, not just economic development. "China will not cooperate, even if the Mongolians ask. And the Mongolians will never ask."

"Wouldn't closing the border be more difficult than collecting legal tariffs?" I inquire.

"We think not. Closing the border is not a matter of discussion with the Chinese. It is up to the Mongolian army to close the border. The Chinese will not voluntarily pay tariffs. No one can collect tariffs," he explains.

"Well, that is a question for the Mongolians. I think it is fair to say that we have enough questions for a fruitful summit of all parties interested in the cashmere trade," I say with feigned optimism. "Can we work together on setting up such a meeting?"

"We are most interested. Draw up some notes and come back," Hara suggests.

I am satisfied. The Japanese have agreed to meet again, and I can guide the discussion by proposing an agenda. That is good for now. It is too early to talk about budgets for the event. I am, however, curious about the accuracy of the Japanese assessment of Mongolia's willingness to confront China on the question of paying tariffs on goods moving across their border from Mongolia.

An opportunity to find out is presenting itself. Two days later, I fly to Hovd, the central town in Hovd aimag, on the remote western edge of Mongolia in the Altai Mountains. It is not far from the geographic meeting point of Mongolia, China, and Russia. I am sitting next to Mr. Davaasambuu, Deputy Director of Foreign Trade and Economic Cooperation for the government of Mongolia.

"You work for the Gobi Initiative?" he asks.

"Yes, we are considering bringing our pilot herders to U.B. to meet with government officials and fiber processors. What would be the prospects of Mongolia approaching the Chinese about respecting Mongolian tariffs on raw cashmere going into China and cracking down on smuggling?"

"Well," he begins hesitantly, "I just returned from Beijing after five years representing Mongolia on trade matters. The Chinese can buy cheap cashmere from our Gobi herders, and they are also seeing the effects of the dzud on Inner Mongolia. They are evicting herders from their grasslands and forcibly moving them to villages where they must become dairy farmers. They know nothing about cows, but you can't fight the Chinese government. Mongolia is in no position to pressure China on anything. Our country of two and a half million people is the size of a small town in rural China. Since the collapse of the Soviet Union, we depend on China for everything. Our only access to a shipping port on the Pacific is one train that goes through Beijing. The Chinese view us as their suppliers of raw materials, not as trading partners. They subsidize their cashmere processors. How can we pressure such a country?"

"If we bring the herders to U.B. to meet the local Mongolian processors, they may make some good business contacts and start doing business with other Mongolians," I say.

"Bring them. The contacts with the Mongolian processors will be useful, but don't expect anything from the government concerning China."

On my return to the office, I call a meeting of the agricultural program staff. "Alta, any luck in finding hay in the Gobi? If not, we need the price of a truckload of hay delivered to Chuulunbaatar. But first, we need to get China into the World Trade Organization by convincing Japan that they do not need their monopoly on Mongolian cashmere for their garment industry, which is competing with China for the world market in sweaters. Let's start by getting all our pilot herders to U.B."

"Of course, boss. The herders have never been to U.B.," says Alta. "Some of their kids, like me, have come, but not the herders. When you were gone, I found hay here in UB. It will cost about two thousand dollars for a truck and trailer, crew, gasoline, and lodging, and another two thousand for the hay. I figure two truck-trailer caravans for each of our three herders. So, eight thousand dollars for each pilot herder; three herders; for about twenty-four thousand dollars altogether."

"Write that up. I'll email Ambassador La Porte as soon as you have it."

"Just one question, boss," says Alta. "Do you think all this is going to work? Will it save Chuulunbaatar's herd and keep him in the program?"

"You know the answer better than I do."

I feel trapped in the city. Although it is close to the government, processors, and traders, it is disconnected from the countryside. The distance is not that far, but reaching people takes a long time, and there is no helpful communication system.

10
In Ulaanbaatar, Mongolia: The legacy of Soviet architecture

The fingerprint of Soviet domination is ostensibly invisible but pervasive in the Gobi Desert. Most of the old wells are covered with sand. Buddhist temples—thousands of them—are ashes in the desert. The Soviets introduced coarse long hair into the herd genetics during the Soviet era. As it grows on goats, it is difficult for the herders to see it.

In the city, however, Soviet influence is tangible.

Figure 17 Downtown Ulaanbaatar

The entrance to my Soviet-era concrete apartment building is an unadorned metal door with no stoop, entrance, or windows. Double steel doors are next to the front door, opening to the bottom of the garbage chute, which extends to the top floor. The double steel doors are usually open, probably from people checking the debris for anything valuable, where the accumulated garbage oozes out of the building, eventually blocking the sidewalk entrance to the front door. During the week, someone comes around with a horse-drawn cart and removes most of the garbage, but the gooey, stinky stuff that sticks to the floor remains forever, so the entrance to each apartment building has its distinctive aroma. There are rows and rows of identical buildings. I can find my apartment building by the aroma when I come home late from the office or after an evening of Russian vodka and German beer.

That method, however, could be more reliable. I prefer the counting method, where I stand by a lamppost with which I have become familiar and count four buildings to the east. A unique collection of posters announcing wrestling matches denotes my lamppost. Mongolians are world-famous wrestlers and have many international champions. The main wrestling stadium is just behind my row of apartment buildings. I have no fear that anyone will ever clean off my lamppost.

The greatest transportation challenge in Mongolia, more daring than driving across the Gobi Desert, is taking an elevator in my building. The elevator is large enough for two passengers or one person with luggage or shopping bags. It will stop at your floor, sometimes giving or taking three or four inches of being level with the floor. But once a day, it stops between floors. Then you yell for the building supervisor or his wife, who lives on the ground floor beside the garbage chute. If they are home, one will flip a switch in the control box to force the elevator to return to the ground floor.

Living on the ninth floor requires using the elevator to haul luggage or heavy shopping bags. I am standing in front of the open elevator, losing my nerve about taking the ride myself, so I place the large box I am hauling into the elevator, push the button for my floor, step back as the elevator door closes, and start running for the stairs. I leap up the stairs to the ninth floor, racing the elevator two or three steps at a time. I lose. The elevator door is closing on my box just as I arrive. My box returns to the ground floor, so I walk down, enter the elevator, and take the challenge of riding the elevator with my cargo. We make it to the ninth floor this time.

This morning, I am off early. I have not been to the office during business hours for many days, so I am anxious to get there. The air is cold and still. My breath is a cloud in front of my face. A haze of coal smoke and heating oil exhaust hangs in the air. I associate the coal and oil smell with frigid, windless days such as this. It is the winter smell of the city, which is colder and darker than the Gobi Desert. City cold pushes down on you. The soot, charcoal, exhaust haze, and the smell of damp cement make the cold unpleasant.

I walk to the busy road next to the wrestling stadium. I stick my arm straight out, parallel to the ground, with my hand flat, palm facing down, and fingers together. Putting your thumb up would be rude, almost like giving the finger. If my palm faced the traffic, it would indicate an order to stop. A palm facing down is a polite request for a ride. Two cars pull up almost immediately. As many buildings have no numbered addresses, you direct your private taxi simply by saying right, left, or straight ahead at intersections until you get where you want to go, then say, 'Stop, please.' I negotiate a modest fee for the service. There are no taxis inUB. This whole process is voluntary by private people who wish to pick up a few dollars for gas. I have never been overcharged or waited more than a few minutes for a ride.

As I walk into the office, Alta greets me. "Nice of you to drop by."

"And it is nice to see you, too," I respond, wondering about the sarcasm. It is always early enough.

"According to the night security man, Chuulunbaatar's son, Batzorig, has been here since early morning. He is looking for money to buy hay," says Alta. "I didn't know what to tell him."

"He is here already?"

"Of course," says Alta in her caustic response, "and he is sitting at your desk now, waiting for you. I told him to try out your computer." The corner of her mouth grins.

I peer into my office. Batzorig is sitting at my desk, his vast shoulders looking even broader with the padded development he is wearing.

Standing beside me, Alta says, "With Layton's help through the Gobi News, we sent all of our contacts a message asking if anyone knows where hay is. No one knows of any, but according to Vance, they either don't know or are lying. It doesn't prove that there is no hay; it only proves that we can't find it."

"Well, of course. You can never prove that there isn't any," I say.

"I understand," says Alta. "USAID just doesn't want to buy any."

"I think you have it. Let's talk to Batzorig in the conference room," I say, walking in that direction.

"Sorry, boss. Kris and Janice are still asleep."

"Who?"

"Our goat consultants from the U.S. arrived last night. They are asleep in the conference room."

"They are two days early," I say.

"Their gear and backpacks are in there, too," says Alta. "Vance wants them out of here. He says they smell like goats, and their clothes are dirty."

They are goat ranchers and have been traveling for about twenty-four hours. Of course, they are dirty, I think to myself.

"Vance thinks that people are supposed to dress up in the office," says Alta.

"They are not going to work in the office. They are going to work in the Gobi Desert. Of course, Kris and Janice are going to smell like goats. They are herders too, just like Chuulunbaatar," I say.

"Boss," says Alta, her eyes looking over my shoulder.

"What?"

"Behind you."

Batzorig is standing a few feet from me. He looms next to me, his broad shoulders even larger. His frame is tilted forward, and his eyes are very penetrating. He is staring at me. He shows no expression, annoyance, or anger, just a presence.

"I didn't expect you so soon," I say.

"I left for U.B. the day after you visited our ger," he says. "I am here to pick up the hay."

"I know, of course. The Ambassador has approved the money, but it will take me another day or so to get it. Can we find hay here, in UB?" I ask, turning to Alta.

"He says he has found some," says Alta.

"I just need to know that none is available near you in Dundgov and that you would sign a contract not to sell it to anyone else. It is a rule of our foreign aid office that you cannot resell anything purchased with U.S. money. Is that OK?"

Alta explains all of this to him. There is much discussion following her explanation. Alta eventually turns to me and says, "He agrees, but he does not understand why we are asking if there is any hay in the Gobi. If there were, he would not be asking us for help."

"What will I say, Alta, that Americans don't trust anyone?"

"Well, that is what I already said to him," she says.

"Excuse me, sir," says our receptionist. "There is a young lady here to see you. She says she is from Dundgov and has traveled two full days. She says something about boots. I will show her in."

"But…," I protest, to no effect.

"Hello, welcome," says Alta, recognizing the boot lady carrying a large sack over her shoulder. They exchange extensive greetings in Mongolian. Alta introduces her. "Sarangerel, this is Batzorig. He is a herder we visited before we found you in Dundgovi," says Alta. "And, of course, you know Mr. Parliament."

"It is nice to see you again, Sarangerel, but why did you come to UB. in person? It is hundreds of miles and probably a three-day trip by truck caravan. You could have sent the boots," I ask.

"I was so worried that they were too big. They are the largest boots I have ever made. They look huge to me," the bootmaker says, looking at the floor.

"I have big feet," I say.

"I was worried," she says again, not looking up. "What if they did not fit? How would I know?"

She talks with her head down in a very soft voice. Alta helps with translation.

"Please, sit down. Let's try them on," I say.

She quickly kneels on the floor before me and pulls a wrapped parcel out of a canvas sack. We are standing in the center of the office. She is dressed in a heavy purple de'el with a dark red sash tied around her waist, desert boots, and a scarf. Her clothing would not be unusual in the Gobi, but in an office in U.B., she is noticed and is beginning to draw attention, especially with the conversation about boots. The receptionists, two people from the bookkeeping department, and the entire business development staff, who have an office area adjoining ours, all wearing suits, are now congregated around the entrance to the conference room. I sit in front of Sarangerel, who is now unwrapping two tall, exquisite boots from her package. She is trembling as she holds one for me to try on. I remove my shoes and socks and slip into the first boot. It is lined with lamb's wool and feels lovely and snug but just right. I pull on the second one, stand, and take a few steps around the office. Batzorig is watching with intensity.

"They are beautiful boots," he says. "The best I have ever seen." Sarangerel smiles at the confirmation from Batzorig, a person of authority.

"They fit perfectly," I say. Applause bursts from the circle of congregants. Sarangerel looks at the boots as I walk around her. She blushes and smiles, a grin as wide as her face, her brilliant white teeth glowing against her reddish-brown skin and black hair. She also looks immensely relieved.

"You must make a pair for me," says Batzorig.

The trip to UB will be good for her business.

"So, you are ready for the nomadic life," says a woman at the edge of what is now a crowd. Our goat consultant, Kris stands beside her sidekick, Janice, awake from the commotion.

"Kris and Janice, this is Sarangerel, who makes excellent boots, and Batzorig, the son of one of our pilot herders, Chuulunbaatar," says Alta, startling me with her skill as the consummate diplomat.

"Nice craftsmanship," says Kris, admiring the boots. Sarangerel absorbs the compliments with grace and pride. Turning to Batzorig, Kris says, "How is the winter dzud treating you?"

"It is a disaster," he says, shaking his head. "We are afraid that we will have no animals left when you bring a buck to us this spring. I am here to bring hay to the herd."

"Are you working with your father?" asks Kris, who was here a year ago when the program first interviewed and selected the pilot herders participating in the animal improvement program. Kris met Chuulunbaatar on that visit before I joined the Gobi project team.

"I am now," says Batzorig, "but will soon have a herd of my own. I have a new family. We will split my father's herd."

"Well, we will get to your settlement first," says Kris. "We need to grade the fiber and check the animals for worms, bone structure, and jaw function," she says in a clinical tone. "Then we can cull the weak ones, so you have a strong base."

"If we cull the herd," says Batzorig, "there will not be enough for me and my younger brother, who will also need animals soon."

"Doesn't the herd go to you as the oldest son?" Kris asks.

"All of us must have animals. How else can they survive?" Batzorig asks as if the answer is obvious.

Ignoring the question's implication, Kris says: "There are too many animals on the land now. The herds have overgrazed the desert. Even without dividing the herd, you must make it smaller. Otherwise, none will be strong."

"The animals are all that my father has. They are his wealth, his worth."

"I have heard him say so myself," I say. "They move as the land gives them grass."

"We will look again, but I think, from what I have heard, that the balance between animal and grass is gone, no matter where you move the herd," says Kris.

Batzorig looks at Kris and then at me. He lives on the experience of hundreds of generations of herders. I have none. He has held every animal in his herd next to his body to give warmth and to move their legs in the bitter cold. He has stacked their frozen dead, slaughtered and eaten their elders, drunk their milk, and cooked the meat from their bodies with their dung. The goats and sheep give their wool without protest to make felt for the ger, and he makes his clothing from hides. He and his family are alive because of them. Kris and Janice are standing across from him, listening. The entire office staff is sitting on desks or chairs or leaning against the wall, listening and wishing they were farther away but unable to move. Many of them are from herder families. Batzorig and I stand facing each other. I wear the finest herder boots in Mongolia, but I do not have the right to wear them.

Batzorig turns and moves toward the office door. Alta runs after him, touching his arm before he leaves. "Come back tomorrow," she says. "Please come back tomorrow. We will have money for the hay.

Alta meets Batzorig in the office. She has arranged for a jeep and driver. On the outskirts of U.B. are herders who have become traders in live animals, pelts, wool hides, meat or bones. Because they keep animals over the winter, they are not nomads. They need hay, and because they are traders, they will sell anything, so they sell hay also. Batzorig knows these traders. He brings hides to them from goats or sheep that have died if the hides are good. Sometimes, the value of sheepskin with good wool is better than the long-term value of the wool if you keep the sheep alive and continue to shear it. Batzorig and other herders will bring sheep-skins to the traders for cash.

Hay is in demand because of the dzud. The price is high. They go from trader to trader checking prices. By the middle of the afternoon, they know the price and can offer it to the trader with the best supply.

"I will pick it up," says Batzorig. I know a caravan that is returning empty to Dundgovi in two days. They will take it, and I will ride with them." A person traveling in the Gobi never has to bargain for transportation. If someone is going somewhere, anyone can ride along. It is another part of the country's common property.

11
The pub in the British Embassy

Batzorig has his hay and is on the way to Dundgovi. My boots fit. I hope Kris and Janice are in a hotel taking showers and washing clothes. I am on my way to the British Embassy. In the basement is a replica of a genuine London pub, serving warm English ales and good Scotch on Friday afternoons. Only ex-pats are invited. The reception is the unofficial meeting place of foreigners working in Mongolia, usually with some connection to their respective governments or the United Nations Development Program, or some international non-governmental organization or "NGO" such as CARE or Junior Achievement, or the Asia Foundation, along with the inevitable intermingling of intelligence officers using these groups as fronts.

If any CIA agents are present, they would use an NGO or U.S. trade mission for cover. "I was talking to a cashmere buyer from China last week," I broadcast in a rather loud voice while talking to a friend from the United Nations Development Program, which has offices in UB. Five heads snap around to look at me. I do this for fun, to smoke them out. They could wear a little American flag next to the gold button microphone on their lapel.

I am talking with Mike Martin, an acquaintance from the International Monetary Fund who frequently travels to Mongolia.

"How do Americans present themselves to the rest of the world?" I ask as we sip Guinness. "What do we really have to say?"

"The American message here in Mongolia," he says, "is that they will spend just enough money to keep their attention and that if Mongolians work with the Americans, more is coming. That may or may not be accurate, but the Mongolians don't know. The other message is that the U.S. wants to keep an eye on China and Russia. Here we are, between two major superpowers, with the China border to the south and the Siberian-Russian border to the north, two of the longest demilitarized borders in the world with a tiny, desperately poor, completely disarmed country in between. The Mongolians are concerned and, at times, suspicious of the Russians and the Chinese, but must engage with both for commercial trade and transportation reasons. Mongolia's natural resources are essential to both countries, as Mongolia is trying to maintain its independence. It is a delicate balance for a fragile state between two world powers.

"So why are Germany and Japan here?" I ask.

"Europe to the west, Russia up north, Japan to the east, and China everywhere else. They are watching each other. So, we all socialize at the pub. It is easier to watch people if they are all in the same place," says my friend.

"But why are they all in Mongolia?"

"This place may hold the balance of power for the world's future," Mike explains.

"That is a bit dramatic. What is here? Silver, no oil, some wool, and fine cashmere," I say. "They do have some oil; they have lots of gold and rare-earth metals," I say.

"The most important thing to all of them, the one thing they will all need when oil is gone in ten years." He takes a sip of Guinness and looks at me more seriously. "The Russians are negotiating for an exclusive uranium exploration deal, and a Canadian mining company is finding significant coal and gold deposits in the southern Gobi! However, their interest is more than just uranium. Associated with the geology of uranium deposits is the likely presence of rare earth metals that are necessary for the manufacture of highly efficient electric generators that are necessary for the conversion from fossil fuel engines to electric vehicles," as Mike explains. Such minerals have yet to be discovered in many places, and Mongolia could hold the key to the future of energy efficiency worldwide.

"That would attract Japan, China, and the U.S. By the way, do you fancy single malt, my friend? On me," I say.

"Most appropriate."

On my way to the bar, a distinguished woman standing near the bar says to me as I pass: "Mr. Parliament, it's a pleasure. Thank you for coming." Kay Coombs, Her Majesty's Ambassador from Great Britain to Mongolia.

"The pleasure is entirely mine," I respond. "I look forward to this gathering every week."

"Well, with a name like yours, you are nearly an official guest," she says.

"I'm not sure the U.S. government would see it that way," I say.

"We always represent our governments, even if we do not intend to, but Friday evening here is for friends, not for governments," she says.

Nothing is personal, especially with the information I just learned. All relations are diplomatic when you are the guest of an Ambassador.

The Ambassador turns to a young woman standing next to her. "Debra, let me introduce you to Mr. Stephen Parliament. He is working with the Gobi Initiative on cashmere herders through USAID. This is very interesting work, especially to our Scottish cashmere wool importers."

"Hello," she says, extending a hand. "Debra Rasmussen with Agriteam Canada, you know, part of the Commonwealth and all," she says as we vigorously shake hands. I expect her to break out in a 'jolly good' something.

"I am also an agriculturalist with goats and sheep. What interests you?"

"Agricultural economics," she says. "We are working on hard wheat that we grow in northern Canada, you know, semi-arid climates like Saskatchewan which is mostly desert, not unlike here. We are helping the Mongolians develop a line of grain that will grow here."

"That would be phenomenal," I say. "I am working with cashmere goats. There is no grass. The dzud is destroying the little grass that there is."

"Exactly. A hard wheat growing here would be beneficial, but more for humans than animals."

"There's nothing wrong with helping humans," I say with a smile. Debra acknowledges the comment. "Let me introduce you to Ingrid," she says as another woman walks past on her way very deliberately toward the bar. Ingrid, this is Mr. Parliament from the Gobi Initiative. He is working with nomadic cashmere goat herders."

"Nomads in the Gobi sounds romantic," she says in Canadian English. "I work for the Canadian Co-operative Association. Nice to meet you," as she hardly slows to the bar. Debra and I move in that direction, too. The bar is two or three people deep. "Gordon," says Debra, noticing a friend standing beside us. "Debra," he says, "how is life in cold old Canada," he says.

"Boring and charming," she says. "More fun here."

"And are you milking any cows yet?" Debra asks.

"Not yet, but we are working on it," says Gordon. "Hi, Gordon Winward, Auckland, New Zealand," he says, introducing himself and shaking hands. "We are looking at the prospects for a dairy industry here in Mongolia. Having good grain will help," he says, smiling at Debra.

"Wouldn't dairy farming tend to reduce grazing land that the herder's need?" I ask.

"It is a big country," says Gordon. "All we need are some wells," says Gordon.

The water that runs south from the voluminous snows of Siberia runs under the Mongolian cap, which has been pushed up to six or seven thousand feet, creating an arid plain. All the water in Russian Siberia runs under Mongolia. Deep wells, yes, but nonetheless, wells can end nomadic life.

"Kiwi Milk Production," says Gordon, handing me his card, which refers to his credentials as a B.S. in chemical engineering and a diploma in technology, specializing in "fat products."

"And here is the money," says Gordon, turning to Mike Martin, who has been listening to our conversation

Yes, they know each other. Canadian demanding wheat growers and New Zealand milk producers need money to bring water to the surface, not for goats. Of course, they would know the International Monetary Fund representative here. I need a beer.

"Can we put some money into short-draft wells for herders?" I ask Mike. "The cost of restoring those old Soviet wells would be minimal."

I could see his interest fade. "Send me a note. The mining companies, wheat growers, and dairy farmers all want water and grassland, which the herders need," he says. "For the herders, you would need thousands of wells all over the Gobi. Drilling is expensive, and then there are holding tanks and windmills. The real expense is long-term maintenance. Who is going to keep them going as the herders migrate?"

"For dairy and mining, the companies could," says Gordon.

"And they could repay IMF loans, but I am not sure how that would work with the herders," thought Mike. "It may have to be the Mongolian government that does it."

"Water wells are certainly not in the USAID budget," I say.

Ambassador Coombs wishes me a cheerful farewell as I leave, looking forward to curling up with my shortwave and the BBC.

12
The herder Tömör

After last week's chaos, the office is quiet, almost sleepy. Ariuna's desk is the congregation point for Mongolian staff. She has a wide socio-electromagnetic field. When we are in U.B., a mini-staff meeting forms around her desk every morning.

"Good morning," she says as I approach. Her voice sounds like she is constantly laughing, under her breath, not out loud. "Yes, and to you," I say. "And what country are we plotting to take over today?" I ask.

"Not China. Too many people," she says. "Maybe Siberia. It is close, no one lives there, and it has water, oil, and lumber. But before that, we need to have a party, a really big party," she says. "Remember, it is part of our work plan. Before you came, we decided to have everyone in the country come together to talk about cashmere."

"I saw that in our work plan. What did you call it, the Cashmere Summit?" I ask.

"Right. Herders, processors, exporters, Mongolian government officials, diplomats, the Japanese, Americans and Australians, and maybe some bankers," she suggests. "And the International Monetary Fund because they have money. And the nice people from the British Embassy. They know how to have a good time," she says in a way that makes me think she knows.

Such a meeting could either be a public relations coup for the United States Ambassador or a substantial national argument. For the first time, the herders, processors, and traders are all in the same place to rationally discuss their interests face-to-face. Whatever is in the interest of one of them is not in the interest of the other two, so the arithmetic doesn't work. But if they don't work together, no one benefits, so they must work together.

"The Mongolian processors want to close the border with China to stop the cashmere from going south," Ariuna says earnestly.

"Yes, but USAID and the U.S. Ambassador do not want that. It would be a restraint of trade, and we are trying to get China into the World Trade Organization. We can't close them off," I say, lecturing.

"Well, get them all together and let them decide," Ariuna says.

"Crazy," mutters Alta to no one in particular, who sits near Ariuna but usually pays no attention to her social circle. Alta stands at her desk and faces us. "The Mongolian processors want the same things as the Japanese: cheap fiber," she says, almost yelling, her arms extended at both sides, imploring. "They want to close the border to keep Chinese competition away so that they can get cashmere and camel fiber. We are an animal improvement program," she says, emphasizing "animal", "not a business development program. Our job is to get the herders a fair price. Japanese investors own many of the Mongolian processing companies. They only want cheap cashmere and camel hair, just like the Chinese."

"Then you can organize the herders and get them to come," says Ariuna enthusiastically, turning to Alta.

"I will have nothing to do with it," says Alta. She turns away from us and sits.

Ariuna smiles and laughs without making any noise. She knows what she must do: organize Mongolia's biggest party ever.

Ariuna and I present the idea to Vance. A cashmere summit is indeed part of our work plan. Still, we have so much to do with animal selection, breeding, nutrition, and forming cooperatives among the herders for marketing that having a national policy summit might not be timely. Ariuna's enthusiasm is highly motivating, and I admit it would be a complicated venture.

"If we are going to do this," says Vance, "then spend some money. Get all the players and stakeholders in the same room. We don't want anyone left out. Make an invitation list for me by EOB tomorrow." He spins his chair so he is facing his computer. The conversation is over.

Ariuna and I invited our three pilot herders and asked them to suggest any neighbors or family members we should invite. By 5:00 pm the next day, I deliver the invitation list. It includes almost three hundred people, depending on how many herders we can find, convince to come, and arrange transportation for. Given other commitments to travel in our schedule, we have three weeks to pull this off. Ariuna dives into the assignment. She rents the largest hall in the largest hotel in U.B. She arranges for a reception on Friday evening with a huge buffet, airag, and live music.

Because the trip from the Gobi to UB is so long, many of the herders arrive a day early. On Friday afternoon, I am taking some of them to retail stores in UB that feature cashmere products made by Mongolian companies.

To my astonishment, many herders had never been to UB I arranged for them to meet Mana, the owner of a processing plant. Five or six herders accompanied me to the plant site on the outskirts of UB.

Mana is a shy, soft-spoken woman of indeterminate age. She greets us in a reception room inside a bleak industrial building. "I appreciate your coming here," she says to our group, introducing herself as an operations manager. I wish I could travel to your homes more often. I have not been to the country in years," she says apologetically. I know your life is complicated. Please tell me how the dzud is treating you.

"We are losing hundreds of animals," one of the herders says. "We are also concerned about the spring fiber production."

"Do you think your company and others in UB can purchase more fiber from Mongolian herders? Now they sell much of it to the Chinese," I ask.

"We certainly can and want to. It is a matter of shipping our finished products out of Mongolia to the world market. It is so hard to get our goods out. We can either send products by plane, which is expensive, or by train through China, but as you know, the train route is long and cumbersome, and the Chinese add fees because they don't want us to compete with them," she says.

"If the herders knew you could sell a specific amount here in UB, they would hold that from the Chinese traders. And you would see that you would have a reliable supply," I say.

"That is good with us," says a herder, "if we have any fiber."

"If several herders pooled their fiber, you could sell it together like a marketing cooperative to Mana," I say.

"That would be fine with me," says Mana.

"We will talk more as the winter goes on," says the herder after talking with others in our group. They must consider the price they might get in U.B. compared to selling to the Chinese. Establishing these connections is the cashmere summit's principal goal.

The summit would be the social event of the year, the first time such an event has ever happened in Mongolia with herders, processors, traders, government officials, diplomats, including the U.S., Japanese, and British Ambassadors, journalists, and translators, all in a huge hall, meandering around with each other. Servers in white coats pour airag, water, and coffee, and roll out food, cart after cart.

Tömör, a distinguished elder from a large herding clan in the South Gobi, close to the border with China, sits at a conference table, surrounded by his sons and other herders. His face is weathered by the burning sun, bone-cracking cold, wind that carries fine sand in the summer, sand and ice crystals in the winter, and smoke. He covers his left eye with a patch. He is tall for a Mongolian, nearly six feet, broad-shouldered, with a one-eyed stare that splits your head in two if he fixes it on you.

An official from the United Nations Development Program is sitting across from Tömör. "Do you think there are too many animals for the land to hold?" the development official asks.

"And which ones will you get rid of? Mine?" Tömör asks, leaning forward slightly, raising his chin a barely noticeable half inch, sending his laser beam single-eyed stare to a spot between the official's two eyes. I know the effect because Tömör did this to me at our first meeting as he arrived in UB. He is looking directly at me. Behind Tömör, his first son stands as tall as Tömör. Behind them, I envision four horsemen, their horses breathing hard in this sub-zero wind, with steam from their flaring nostrils. They hold staff straight into the air, tying colored ribbons to the staff, and wear dells of purple, gold, and dark red. The horses are impatient to move on as their sweat freezes on their coats. Tömör and his sons hold all the power of the Mongolian tradition of dominance in their eyes and countenance. They are compelling.

"We do not kill animals. We do not trade them. We move them to where they are needed," Tömör says. Tömör rises, takes a cup of airag in his hand, places his index finger into it, flipping liquid in front, behind, and to each side, and, saying nothing, lifts the cup to the U.N. development official sitting in front of him. The official finds a cup, many of which are on the table, and raises it while standing in acknowledgment. Tömör pulls out his tobacco jar from inside his sleeve and places it in the open, upward-pointing palm of his right hand. At the same time, he holds his right elbow with his left hand and extends the tobacco to the development official, who takes it, pinches tobacco into his nose, and returns the jar.

The U.N. development official steps aside as Tömör and his sons walk into the center of the ballroom in the Genghis Hotel. The processors will implore the Mongolian government to close the border with China for the next two days. The next day, as the arguments continue without a consensus, Vance walks past me in the hallway.

"You might as well close this," he says. "All the important people have left. The resolutions they are working on do nothing. Take the herders out to dinner."

The herders are outnumbered, and the commercial interests in UB. have more influence here. An opportunity for a cooperative consensus is lost. The herders must be better organized to defend their interests more effectively.

border with China for the next two days. The next day, as the

you reach a waterfall in the...the important people

13
A quiet woman

A few days after the summit, I come to the office early. Alta is already sitting at her desk, looking out the window. She does not turn to see who I am as I enter our office.

"You are here early," I say. No response. "The herders have gone home, but I think we made some useful contacts for them with the processors in U.B." Her silence is not unusual, so I do not place meaning in it. My desk has letters, memos, and office stuff piled on it, some bills that have arrived at the Gobi Initiative office that we need to pay from the special funds set aside for the summit, and a statement from the transporter who took the hay to Chuulunbaatar, who was not able to come to the summit conference.

"Looks like we got the hay to Chuulunbaatar. I have an invoice here," I say.

"That's good," says Alta, still not turning around.

"We need to find some good, elite bucks for our herders."

"If there are any does left," says Alta.

"What about Udval? Is she in U.B.?" I ask.

"She traveled with some herders, but Chuulunbaatar could not make the trip. He had to stay with the herd."

"Is she in a hospital?" I ask.

"Why do we wait so long?" asks Alta, who is not responding to my question. "Why wait for the sheep and goats to die before we get hay? Why wait for the does to die before we bring strong bucks? Why did we wait for Udval to come here instead of sending a doctor to her? We knew she was sick, or at least we should have known a year ago. Why didn't we know?"

"Alta, what happened?"

"Why don't we know anything?" she says, finally turning in her swivel office chair to look at me, her eyes red and her face moist with tears. "They put her in a caravan truck coming to U.B. They stopped on a mountain pass to put stones on the pile to ensure a safe trip and that they would return. She did not get out of the back seat. They looked in, and she was curled up, hunched over like she is when she walks in pain. But she couldn't get up. Now, I can't find her, and I cannot find out anything about her. I don't know if she got to a hospital. Did we do what we promised? Why do we wait?"

At the moment, I have no answers for her. I go home and slip off my boots.

14
The culling

Our caravan leaves UB early in the morning. We will look for a high-quality elite buck, find Chuulunbaatar, and ensure he receives the hay we sent. Then, we will talk about culling his weak animals.

We head southwest to Mandalgovi, an aimag center in Dundgovi, where a local breeder has some good bucks. Our information comes from a reliable source: a veterinarian named Alice, who volunteers her time through Agricultural Cooperative Development International. They send professional volunteers on short-term assignments to developing countries, often associated with a USAID-funded project like ours.

Alice consults with cashmere raisers in the United States and Scotland and has visited Australia, also known for its good quality cashmere. But for her, the finest cashmere in the world is in Mongolia, which arrived here from its original home in Kashmir. Buddhist monks from Kashmir and Tibet brought gifts of cashmere goats for Khubilai Khan in Mongolia, where Buddhism was taking hold. One of the most remarkable attributes of the Mongolian empire between the thirteenth and fifteenth centuries was the tolerance of religious diversity and the significant expansion of trade. Goods and religious beliefs moved in all directions across Asia to Europe and back, facilitated and encouraged by the Mongolians.

The Tibetan monk's cashmere gift to the Khans has had a profound and lasting effect on life in the Mongolian high plateau. The world's best cashmere still comes from either Inner Mongolia in China or Mongolia. The goats adapted very well to the harsh winters of the steppe. The coarse guard hair keeps the dirt and snow off, but the fine inner down keeps them warm.

Though tough enough to thrive in a highly arid climate, they are threatened by intestinal worms and genetic defects that sometimes take a few generations to reveal themselves. Alice, our veterinarian friend, taught a deworming technique to one of our animal consultants, who now goes with us on trips to the herder encampments to teach it to the herders' families. Alice will kneel next to a goat and place her arms around its neck, sliding her hands to the back of the head and lower jaw. She pushes a hand between the teeth of the goat, like slipping a harness into a horse's mouth so that the mouth opens just enough to insert a small plastic syringe with deworming medicine. A quick push on the plunger is all that is required. The process must be repeated two or three times over a month to ensure success.

The genetic problems are more challenging to correct.

The first genetic difficulty arose during the Soviet period, which began in 1924 with the formation of the Mongolian People's Republic. The Soviet grand design for Mongolia emphasized cashmere quantity over quality. The Russians introduced goats with much coarser hair to increase fiber volume. The long guard hair slowly displaced the fine down, and it is now disappearing.

The comparison of fiber from a high-quality goat with that of a coarse-haired goat is startling and discouraging. The only way to fix the problem is to remove the coarse-haired goats and start a breeding cycle with elite bucks. It will take time. Fortunately, Mongolia has high-quality, elite bucks so herders do not have to rely on bucks from other countries. The challenge is finding them and making them available in the Gobi.

The other genetic problems include teeth that don't line up, preventing goats from eating properly, weak chests, crooked backbones that make it hard for a female goat or doe to carry a heavy load while pregnant or with udders full of milk, and hindquarters that are so narrow that the female cannot easily give birth.

As I talk to the herders about culling their herd, the words "cull" and "kill" are the same in Mongolian. Instead of using the term "cull," my team talked about "improving" the herd's genetics selectively. We are doing the latter. The herders hear the former.

To give a more positive spin to what we are trying to accomplish, we have a trainer with us, the son of a herder family. His name is Temüjin. He is taking classes in business and biology at the Mongolian University of Science and Technology in UB. He is very knowledgeable about the practical art of sorting high-quality fiber. He uses camel, sheep, and cashmere goat fiber samples to show how to detect different grades. We aim to show the tremendous advantage of sorting fine fiber from coarse before going to market.

As we drive closer to Mandalgovi', we pass clusters of three or four gers with a rough wooden fence surrounding them. These are usually residences for retired herders taking meat, hides, and vegetables to the local market. Some are repair shops for old vehicles or stores selling reconditioned old clothing. In the center of town is the post office, where they hold mail until residents or nomads come to collect it. The people in the post office know everyone, so we can find the breeder and his elite bucks.

Selecting what appears to be the best buck available, Temüjin begins his inspection. He opens the goat's mouth, checking teeth and jaw alignment, the chest for size and strength, and the hindquarters. Temüjin combs the buck's hair, pulling some underbelly fiber. He carefully holds up a handful of cashmere, gently separating it and turning it to sunlight. It is white and so soft that it appears to be weightless.

"This is the most beautiful cashmere I have ever seen," says Temüjin in a quiet, respectful voice. "We should buy this buck and any others like him."

We have an extra truck for this purpose, so we purchase three elite bucks at more than a fair price. We will take them to Chuulunbaatar, Tömör, and Dembereldorj, the three herders who have agreed to work with us on this experiment to improve the quality of cashmere fiber.

Many families in the aimag centers have a spare ger or two available to rent—Mongolia's version of a motel. We rent a large ger for the night. Above us, the Big Dipper and North Star are overhead instead of on the horizon, which I am accustomed to seeing. The aimag center has almost no nighttime lights, and the sky has no clouds, no distorting pollution, no heat waves, nothing to interfere with the sublime, dark black sky and dazzlingly bright starlight.

We each have a massive pile of felt blankets on a mat arranged around the inside circumference of the ger. After we climb underneath, a small fire in the cooking stove continues for a few minutes. Once in bed, the fire is not needed to keep us warm.

We loaded the three elite bucks into our truck the following day and headed for Chuulunbaatar's. It took two days to find him. The greeting was warm and long. We took the bucks to a special pen to be cared for.

His whole family is there, and an extra ger for visitors.

"I am so sorry about the loss of your wife," I say.

"My daughter and sons are caring for me," he says.

As Alta translates my sentiment, everyone in the ger stops talking. I suddenly do not know if it is appropriate to recognize the death of a spouse, but sorrow for him is what I feel, so it is honest.

"It is what you feel, so it is OK," says Alta, sensing that I am unsure if I said the wrong thing.

After the airag and tobacco ceremonies, we discuss the buck and how to identify the best cashmere fiber for the market.

Temüjin stands in the center of the ger, with herders, colleagues, and children sitting on the floor surrounding him. He is proud of the attention and respect he deserves. Though a young man, his audience senses his seriousness. Everyone is quiet.

Temüjin removes a dozen clear plastic baggies from his shoulder bag and places them on the floor of the ger in a circle. One of the kids runs outside to find a few missing brothers and sisters. He must know that this will be a performance not to miss. He adds camel dung to the stove. A large steel bowl of brick tea is being made, filling the space with the beautiful smell of fire, tea, powdered milk, steam, and the warm felt comprising the ger's interior walls. Soon, over twenty-five people are watching Temüjin. Alta and I take a baggie together. I wonder if I can tell the difference between camel, sheep, and cashmere fiber. How embarrassing it will be if I can't?

Temüjin opens his larger plastic bag and pulls out nine small clear plastic baggies, each one with a handful of some fiber inside and a piece of an index card with a label written on it. He holds one of the small baggies. He has very fine, medium, and coarse hair samples for each animal. "This hair is unwashed, uncombed, just the way it comes off the animal," he says, instructing us to sort the bags, first by separating those with coarse hair from fine and then by sheep, camel, and goat. Everyone scurries to their baggies. There is a lot of chatter, laughter, and arguing over which is which. Even though they are labeled, no one is paying any attention to the labels. They want to sort for themselves, including Alta and me. Temüjn looks at me and smiles. The interaction could not be going better. We were unsure if the herders would be receptive, but they love this, especially the kids.

Temüjin is an encouraging and enthusiastic teacher. He wants us to compare and see the differences before he tells us. All he says is, "Don't mix one fiber with another. Then we would have to throw out that batch." After everyone has sorted, touched, pulled fiber apart, and lumped it together again, he checks our work to see if we have managed to distinguish among the grades and then by sheep, goat, and camel. The camel fiber is relatively easy because of its distinct color, but sorting fine goat fiber from sheep wool is difficult. Sheep's wool will be primarily white with a little black, but there are more critical differences than color.

"Smell the fiber, and when you press it on your nose and hold it tight, you can feel the lanolin."

Everyone presses a handful of wool to their nose and face.

"Now take the goat hair. What's the difference?"

There are many theories, but what it comes down to is "fluffier." The cashmere is lighter. It floats in your hand. The difference is nearly impossible to see with your eyes. Cashmere fiber is very narrow, with little crinkles in it. High-quality cashmere is less than 17 microns or 17 millionths of a meter in diameter. For comparison, human hair is between 57 and 90 microns thick, and Asian human hair, like that of the people in the ger, is about 120 microns. The coarse hair is straight so you can pull it apart more easily. The best fine hair holds onto itself. If you could take one strand each of camel, wool, and cashmere fiber, cut it, and look at the cross-section, the camel would be the thickest, followed by wool and then cashmere would be much thinner. The fibers hold onto each other, making it easy to spin into yarn. All the fine fibers are like that, but cashmere is the thinnest, lightest, and has the most crinkles.

"The processors know the differences," says Temüjin as he examines our work. "If you don't want to get cheated, you had better know also."

Temüjin explains to Chuulunbaatar that we have brought an excellent elite buck to give him the best possible fiber, holding up the fine white cashmere sample. Use t is a buck, he says, and three years from now, you will have the finest herd in Dundgovi.

Near the door, our boots and coats are piled up. My boots are made like the others: dark horsehide, tall, with flat, thick soles and Mongolian symbols embossed on the side. As I reach for them, I realize they are enormous, so much larger than all the others that they must look like the boots of a giant. I feel so much a part of these lives and this place that my perception of difference has diminished, but maybe that is an illusion. To them, I must still be a foreigner, a visitor at best, a friend I would hope, but I realize that I am a friend with money, with the ability to buy a good buck for them and hay in the winter.

The bright sun on a cold, clear day is always shocking to the eyes. It takes a minute to adjust from inside the ger. Temüjin walks with pride to the new buck, which is handsome, well-fed, and healthy, with a long, shaggy overcoat that hangs nearly to the ground. The buck is not tall, but maybe four hands to the top of the shoulder at most. Temüjin recommends to Chuulunbaatar that he breed this buck only with the best does, just a few, and watch for results.

We talk about worm medicine and how to apply the ear tags that identify strong does. I give him a device resembling a staple gun to attach the ear tags.

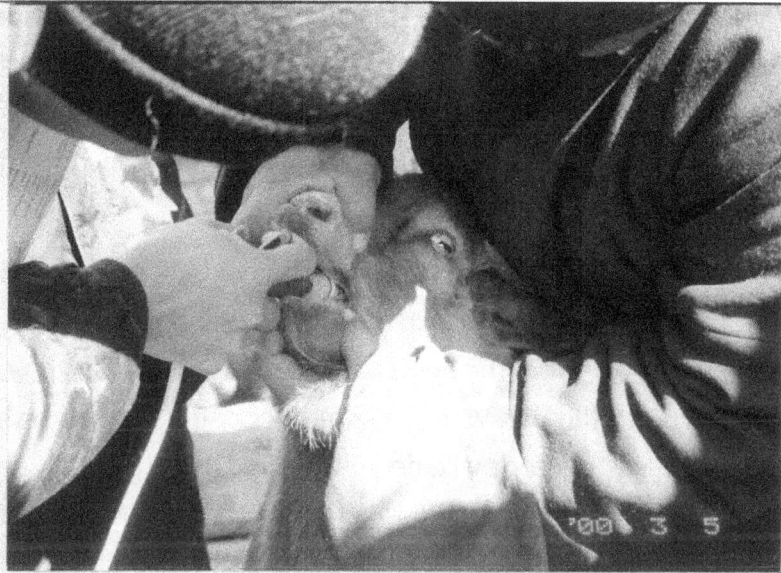

Figure 18 Administering de-worm medicine

We hear the sound of animals over a rise behind the fences. More goats come over the hill, followed by a camel with a rider, clumping along in that ancient ornery disgruntled shuffle that all camels have. The only time they seem to be at peace is when you see a long line of them, miles away, walking along a riverbed, not being driven but moving with a purpose all their own. They won't lie down like cows in a field. They must be in large groups. They are very gregarious animals and cannot survive alone or with just two or three. If you have camels, you must have many of them, so the herders let them wander where they want and find them when needed.

The goats are herded near the fenced area and quickly join those already there. Some have yellow ear tags, and some do not. Alta looks at me without speaking as if to say she is skeptical of the plan. The fencing is very tenuous. Herders do not take the fences with them when they migrate to a new location. They leave them for someone else. It makes it challenging to keep the animals separated for breeding purposes. Alta and I share the concern without speaking.

Temüjin leads the elite buck to an area that looks like it will hold him, showing the oldest son how to apply the ear tags. The son takes a good doe from the herd, wrestling it through the others. He puts it into the pen with the elite buck. "And another," says Chuulunbaatar. The son's wife begins to help. One at a time, they push the best does through the herd and place them into the pen with the buck. It takes an hour or more. Chuulunbaatar is leaning on the fence, on the inside, looking out at us but also at the horizon or at nothing. He is reducing his wealth, prestige, and endowment for his children because his herd will be smaller. One of the family members invites me to ride the camel, an offer I readily accept. The Mongolian model has the distinct advantage of having a natural saddle: no stirrups needed. Despite their cantankerous personality, camels are easy to ride. The Mongolian camel, or "Bactrian," has two humps to store extra fat in the cold northern climate. The single-humped "Dromedary" variety is more widely domesticated and found in North Africa and the Middle East.

The camel kneels to the ground with its legs folded and its head very erect. It complains as I climb on, spitting and snorting in a massive fit of perturbation at having to move. Half the ride's fun is the perverse pleasure of needling such a grumpy fellow. Even in his complaining, he is easy to lead, going in any direction I wish without resistance or the single-mindedness that you sometimes find with horses. In all respects, this is a cultured and refined beast with a stature of dignity, aloofness, and extreme self-centeredness. The younger children and I take turns riding the camel around the encampment. Alta, of course, is disdainful. "Smelly thing. Why would I want to sit on it? You will smell like a camel for three days."

Figure 19 The author on a two-humped "Bactrian" camel

I slide off the camel, running my fingers through the long, coarse hair, not willing to risk a search for the fine hair on the underbelly. If a camel does not know you, pulling hair off its belly might cause the animal to remove a pound of your flesh in exchange. I prefer the ride and will not bother the irritable camel.

As dinner is ready, we sit at the table next to Chuulunbaatar, giving a toast to each other and all those present in the extended family. A pot is boiling vigorously on the stove, and airag is flowing. The ger seems to be rotating in a circle. I feel the camel moving under me like a carousel horse. I am still trying to figure out Chuulunbaatar's intentions regarding the herd, but at least the elite buck is here.

Amid much noise, Chuulunbaatar rises, turning to face Alta. He raises his cup and fills it again with airag, then fills hers. She does not drink. She is barely able to stand after many toasts. "I have a dedication for you," says Chuulunbaatar. He reaches into the boiling pot on the stove and pulls out a hindquarter of goat, dropping the steaming carcass onto the short wooden table beside him. "This time, you have the honor of the first cut," he says, handing Alta his knife. She looks at the rump and leg steaming before her on the table. The first cut of meat is always reserved for the older leader of the visiting party. It is indeed a privilege for Chuulunbaatar to hand me the knife.

Chuulunbaatar raises his cup, indicating that Alta should do the same, which she does. They drink. In a grand flourish, Chuulunbaatar says, "If I must cull does from my herd, killing my livelihood, then you must eat them," gesturing for her now to carve the meal and take the first bite.

He has decided what he must do. We can leave him now with his sons, daughters, and daughters-in-law. He will now cull animals that are not starving to death on their own, something he never thought he would do.

As we walk towards the vehicles, reluctant to go, saying goodbye over and over and reassuring them we will be back, Alta stops, looks at me, and says, "I have only one word for you."

"What?"

She takes a deep breath, steps back a pace, starts running to the trucks with one arm raised, and shouts, "SHOTGUN!" at the top of her lungs so that everyone can hear, jumping into the front seat of the Land Rover to secure her favorite spot.

15
Snow hawks and survival

The Gobi is blue and white as we drive through it, with small bushes, one foot to eighteen inches tall, scraggly with thin leaves that are dead and brown in the winter. The wind blows from the northwest. A mixture of dust and sand cemented together by icy snow builds on the northwest-facing side of the bushes so that by the middle of winter, each bush has a dirty snowdrift leaning on it. The left side of the bush is like a tiny cave one of thousands in the desert.

The sun warms the little caves with its direct light and the reflection of the snowdrift inside, providing a perfect shelter for small hawks, which in the winter turn mostly white for camouflage so they are less visible to their prey. The hawks wait for gophers that have inundated the Gobi, eating roots and ruining vegetation for the goats. They burrow miles of tunnels under the desert's surface, protecting them from the hawks and keeping them warm in the winter.

These tunnels become booby traps for camels, horses, goats, and sheep as they move across the soft desert surface.

The gophers multiply faster than the hawks can capture them, so there is an imbalance. Eventually, the hawks will restore a balance, but it will take time. Nature has time, but the herders have none. The grassland is disappearing. The desert is becoming less habitable for the herds. Even our Land Rover struggles with gopher holes. How much can the desert bear? Every living creature wants part of the desert, but it is fragile.

Will herding become extinct if the gophers, the newly arriving gold miners, or the dzud make life impossible for the herder? Is this just a cycle of survival and challenge, or are we witnessing or even participating in the end of the last ancient nomadic culture on Earth, which has supported these people for centuries?

I have faith in the beauty of these animals, their caretakers, and the exquisite gift they give us through their fur coats. They are worth defending, so we will keep looking for Dembereldorj and help him as much as possible.

The last time Buyan found Dembereldorj was more than a year ago. We head to where he was last seen, in eastern Omnogovi'. He is far from UB, much farther than the other pilot herders in our program. It is essential to USAID that we demonstrate our ability to reach those far away. We are showing more than modern veterinary medicine and cooperative marketing. We have to show a new method of preparing goods for the market. With small loads of high-quality fiber, we can transport products longer distances with fewer vehicles than the enormous piles of hides and raw fiber that are customary.

Traveling and searching take time. I remind myself about the Mongolian quality of patience. Traveling in the Gobi seems motionless. I feel like we are standing still. The valleys and ridges are indistinguishable. There are no distance signs with arrows: Dembereldorj, 25 km; Aimag Center, 45 km; UB, 389 km. You don't know how far away anything is; you only see when you arrive. The Great Khans rode and walked from one end of the known world to the other. The Mongol nation has existed intact since the 1200s when Genghis Khan unified his rule over all the steppe tribes. By 1205, he had conquered Northern China, and in about 1218, he set out on his campaign to the west. After his death, the Mongols advanced through Persia and the Ottoman Empire to Hungary in the south and Russia and Finland to the north. The expedition into Europe halted at Vienna with the death of Ogedei Khan. From 1240 to 1480, the Mongols maintained control over the Russians, who called them the Golden Horde. In 1260, Khubilai Khan became the Great Khan in Beijing, China, where Marco Polo was his diplomatic emissary, taking notes as he went. These would later become a bestseller in Europe and the subject of poetic tribute by Coleridge as Xanadu, or Shangdu to the Chinese.

The Mongolian Yuan dynasty in China lasted from 1279 to 1368. In 1388, the Chinese destroyed and obliterated the beloved Mongolian capital of Karakorum, ending the longest and most geographically extensive empire, the world has ever known. The empire stretched from South Korea to south China, to the Black Sea in Asia Minor, to the White Sea in the far north, and through Ukraine, Belarus, Bulgaria, and Hungary.

Americans are no match for the patience of Mongolians. Our search for Dembereldorj takes many days, and we become impatient. We find small settlements along the road that offer a bowl of soup and a ger to rent for the night. The few hundred people who live in these settlements are not nomads but tradespeople who live off the travelers on the road. Buyan will ask a local where to rent a ger for the night. We always find one. The owner always starts a fire as we enter, but they are freezing inside. Each of us finds a pile of felt and wool blankets around the perimeter of the ger. It is usually late at night when we stop, so the fire is superficial. We burrow under a heavy pile of blankets and fall asleep. The felt's weight and the wool's smell are comforting and reassuring. The smoke from the coals flickers in the stove. The wind rattles the pelts against the ger frame in a soft rhythm.

I have never slept better or warmer than in a ger in the Gobi Desert at twenty below zero, with the wind blowing until it is time to find the outhouse. There is a point at which it no longer makes sense to wait. Two hours later, I leave my felt cave and see the door.

Wood slat fences surround the gers in these settlements to keep out marauding gangs of hungry wild dogs.

Figure 20 Settlement with wooden slat fence for protection

During my first visit to UB a few years ago, I was often awakened by the barking and frantic scurrying of dogs running through the streets of UB at four or five in the morning, howling and fighting as they searched for food. There must be wild dogs here, too, or wolves, as I venture outside with a flashlight but no weapon. Necessity overcomes fear. The outhouse is a three-sided wood frame with a large hole in the ground and one board across it. You place your feet on the board, holding onto the side of the frame for stability. The slats are about one or two inches apart, so the wind whistles through. The night is beautiful, but the waiting pile of felt is much more appealing.

When we find Dembereldorj, we talk about his animals, the impossibility of finding hay anywhere for purchase, and the price of cashmere. Dembereldorj is a capable herder with hundreds of animals, though the pile of dead ones is getting disturbingly large. He told me he would keep working with us if we sent him hay, but he could not come to UB to collect it.

Alta reminds us that there are more severe obstacles to helping him than purchasing hay. Delivering the hay is more complicated. It is the challenge of distance and transportation, not just money. We stayed the next night in his compound, but our group was weary and anxious to start home.

As we travel to UB on the return trip, Alta says that she would like to find a woman named Batsukh who may be able to help us with high-quality elite bucks. Alta heard a story about her on the radio and wanted to meet her. She lives north of UB and has won awards for her herding abilities. We tried to find her, so we took a northern loop around UB to the town mentioned in the radio broadcast and asked for her. She is very well known. It does not take long to find her.

As we arrive, a woman and older children are setting up tables outside their ger, with younger children running in circles. The older children bring steaming pots out of the ger for the younger ones.

Alta hops out of the Land Rover and walks to greet the woman, asking if she is Batsukh, the woman interviewed on the radio. The woman confirms that she is Batsukh and waves at us to join her. Alta explains our program.

"You have many responsibilities," I say, arms stretched out, indicating all the children and animals in the vicinity. She smiles modestly, saying she loves her work and has taken in many children without parents.

Batsukh has a small table with folding chairs in front of the main ger. The older children bring us tea and biscuits. The children are curious about us. They cluster in small groups, getting as close as possible without touching, and then they giggle hysterically while running away.

"Please excuse them," says Batsukh, returning with cups and a teapot. We don't have many Western visitors."

Alta asks her about a comment in the radio interview about conflicts with other herders. She explains that women find life especially hard in the countryside. Other herders always try to take advantage of females, like not allowing them to use certain areas of good grassland.

"We formed the Liberal Women's Brain Pool to support women in the countryside. I know it is a funny name, but it scares the men," she says in a tone that means that scaring men isn't hard to do. "We have to think in the long term for our children. Women will make better politicians for the country than men, who think only of immediate power. That's what Dr. S. Oyun, the leader of our movement, says."

Dr. S. Oyun was the leader of the Democratic Union, which challenged the Mongolian People's Revolutionary Party (MPRP) after Mongolia's independence from the Soviet Union. She is now head of the Real Party, the only member of her party who is also a Member of Parliament. More recently, she was head of the Democratic Party but refused to participate in a coalition that included the MPRP because of its alignment with Russian communists. She is convinced that the Russians were at least indirectly responsible for the assassination of her brother, S. Zorig. His murder has not been solved.

Zorig's sister, Dr. Oyun, says the women's movement is the secret to Mongolia's future. The women are equally capable of sharing herding duties as the men. They can keep herding, as Batsukh does if they find themselves without a husband. We stopped at a ger as we traveled through the Gobi to ask if they knew whether Chuulunbaatar or Dembereldorj were nearby. The adult in the household is often a single woman with older children. They're able to keep herding because they have always shared the work. Alta is quite emphatic about this. "Of course, women throughout the country are just as capable as men," she says, not just about herding. "Sometimes it can be dangerous, but they know what they are doing."

In addition to their domestic equality, women are well educated. With no television in the countryside, the families read. The national literacy rate is 97.8per cent, with almost no distinction between men and women. It is one of the highest in the world, which is a remarkable accomplishment for a country that ranks as the 160th poorest country in the world. However, school enrollment figures are beginning to show a distinction between the sexes. The percentage of age-eligible males enrolled in secondary school is 92.1per cent, while the percentage of females is 95.6per cent. This difference becomes more pronounced at the university level, where women comprise 67per cent of the enrollment in higher education. The national university system is quite distinguished for a small and isolated country.

The most prestigious is the National University of Mongolia, the oldest institution of higher learning, which was once the training center for the Communist Party elite. According to some international university ranking services, the methodologies of which can be subjective, it is ranked 2,402 out of approximately 18,000 colleges and universities worldwide. It is now an internationally recognized university with 12,000 undergraduate and 2,000 graduate students in 80 programs. The Mongolian University of Science and Technology, where Temüjin is enrolled, has 24,000 students and an extensive curriculum with many Ph.D. programs. The country has thirteen colleges altogether.

These schools all need help to obtain up-to-date texts and scientific equipment. English is increasingly the language of instruction, but language training at the secondary level could be stronger. The Mongolian State University of Education is promoting majors in teaching English, with many international instructors and Peace Corps teachers involved. As I wrote a column for the Gobi Business News, which my organization published, we encountered constant difficulties translating technical and medical terms into Mongolian. A colleague and I started working on a Mongolian-English dictionary for technical and business languages. Even words such as net income retained earnings, antibiotics, and animal husbandry took a lot of work to translate. I regret to say that time prevented me from staying with the project. Hopefully, it will be continued.

With more women in higher education, the scientific and international finance requirements of the future will ensure their direct participation in planning. Two of my colleagues at the Gobi Initiative, Itgel, who worked in the business and small banking sector, and of course, Alta, both obtained advanced degrees from major universities outside of Mongolia, returning to take positions of responsibility in the country, Itgel with the Asia Development Bank and Alta with the Millennium Challenge Account. This pattern will continue as long as women remain in secondary school through graduation.

Dr. S. Oyun is a terrific role model for educational advancement. She has a Ph.D. in geology from Cambridge University in England and is a Member of Parliament. Batsukh admires her. Though self-taught, Batsukh has the support of the Women's Information Resource Center and the Gender Center for Sustainable

Development in UB, both of which help Mongolian women start their own businesses and advocate for women's rights.

As we talk with Batsukh, she reminds us that women herders are not treated with the same respect as male herders regarding the right to use grazing land. She has difficulty relocating her herd to a new land, and women are often not allowed into a new grazing area.

Before we leave, we discuss the possibility of purchasing an elite buck from her and whether she might be interested in joining our program for herd improvement. She is more than happy to lend us a good buck that we can use for breeding service, but she wants it back. Regarding the improvement of her herd, she is already culling weak does. She is not caught in the cultural trap of demonstrating wealth by the size of her herd. She understands that the quality of the fiber is what matters.

As we drive toward the city, I start thinking about the coming week in the office. When our Chief of Party Vance and I last met with the U.S. Ambassador to Mongolia, Alphonse LaPorte, we mentioned the possibility of meeting with two Mongolian cooperativists, Erdene and Nadmid. I asked Ariuna to set up appointments for us while we were traveling. I want to ensure I pursue those contacts when we return to UB. Vance was skeptical because these people were collectivists from the Soviet era. However, they still might have some insights about bringing herders together and are respected among the older herders for their advocacy. They may provide a bridge of confidence between us and the herders, who seem skeptical about outside advisors.

I look forward to meeting them.

16
Cooperation among the herders

A few days later, in the office, I ask Ariuna to re-confirm appointments with Erdene and Nadmid, who have both worked with herder associations. I expect them to refrain from promoting the idea of a free market, but they have experience with herders working together on common problems. R. Erdene is President of the Mongolian Association of Private Herders' Cooperatives, of which I know of none. He is also the Executive Director of the Training Centre of Cooperatives, which does exist.

N. Nadmid is the President of the National Association of Mongolian Agricultural Cooperatives. He has formal training in animal science and is a soft-spoken, polite teacher at the agricultural university who is curious about new ideas. Erdene and Nadmid are friends and complement each other perfectly. Erdene is more of a businessman, mechanic, and farmer who believes that small, efficient machinery will ultimately be the herders' salvation if they agree to cultivate instead of wander. Cultivation, however, leads to land ownership instead of open common space. Alta and I have yet to talk about cultivation. Still, we have been promoting cooperative marketing as a way of leveraging a better price for good fiber and transporting goods from the Gobi to UB for sale more efficiently.

Alta and I have appointments scheduled with Erdene and Nadmid, but I want to visit Erdene's training center before meeting with them. It is a single-story cement block structure on the outskirts of UB. The building is very long, with a small door in the middle, entering into a blank foyer with offices on the side. Erdene greets us at the front door. Straight ahead is a room with no furniture, like a high school cafeteria without tables or chairs. On the far side is another door that leads to a machine shop. The shop has two or three small dark green tractors, still partially in wooden shipping frames.

"These are very efficient, easy-to-assemble tractors that a herder can learn to use for garden plots or small fields," says Erdene. "We can show them how to assemble and use them here," he says.

"Do you think the herders are ready for cultivating crops, and how would the tractors be delivered to the nomads in the desert?" I ask.

"They would go by truck caravan to the sum centers and be assembled there. The fields would be near the center," he says. "The fields would be cultivated cooperatively, with each herder family sharing in the work."

"I thought that system did not work too well," I say.

"This is all private, people working together," he says.

Nadmid, who is with us, is silent. I take his silence for disagreement. We go to a conference room in the building. I ask them how many cooperative associations there are now and how many herders belong to them. It is difficult to tell, but certainly fewer than before the end of the Soviet era in 1990, they tell me. "I want to show them what they can do working together," I say. "I want to see if they can work together on a business. Erdene, you showed me a tractor that the herders could put together and share, but it seems they need a good truck to pick up all their fiber and pelts and drive everything to UB for the market. They could do this together, so each family does not need a truck, just like your tractor idea. Do you think we could make it work?"

"Yes, we could. When the government privatized the herds, everyone took on a very individual way of thinking, and we thought that was what private enterprise meant," says Erdene.

"We can talk about something that the herders own instead of the government; the herders began to get the idea. The members own land and equipment as a cooperative business. What I need is a business plan so you can see what I mean," I say. "I am going to take a sheet of graph paper and draw a line down the center. On the left side, I will list everything that a herding family spends money on if you would help me make the list."

Nadmid, Erdene, Alta, and I all start talking, and I take notes. Herders buy tea, sugar, dried milk, vegetables, medicine for themselves and sometimes for animals, clothes, gasoline, knives, rope, and many miscellaneous items. We add up the cost for a month and a year in round numbers. Then, on the right side of the graph paper, I ask them what herders sell. We can estimate what they would make from selling cashmere, sheep's wool, camel hair, sheepskins, horsehide, camel milk, fermented mare's milk as airag, and dried meat. "How much do they sell each month?" It has to balance. A business only works if income and costs are balanced. If your costs exceed your income, you are in trouble." The exercise is complicated and requires more price, revenue, and expense information. To do this right, we need a business accountant and training for their membership.

In our exercise, the left-hand side of costs is more than the right-hand side of income. "So, we increase the price for cashmere that they need," I say. "Not what the Chinese traders offer, but what they need to survive. If they don't get what they need, the processors or the traders are not paying enough. The left and right sides have to balance," I say, holding my hands and moving them up and down like a scale. The higher the quality of the goods, the better the price, which is why we discuss improving the herds' genetics. "This isn't fancy; it is just market economics," I say.

"This is a new way of thinking," says Nadmid. "This is what we need to know. How can we do this?"

"If some of the herders could talk with UB processors, it would be the start of understanding how the market might work," I suggest. Alta recommends meeting with two UB processors to see if we can set up a relationship between processors and herders, like a futures market or an agreement to buy fiber. Maybe the processors would advance some money to help the herders get through the winter if the herders agree to sell to those processors in the spring," says Alta.

"Like the Soviet system of a guarantee to purchase?" asks Nadmid.

"The difference would be that the price has to be negotiable, given the market demand in the spring. The price has to be flexible, not fixed as it was during the Soviet period," I say.

"If the processors could advance the herders enough to buy hay in the winter, the amount they gave would certainly be less than what cashmere will be worth in the spring," says Alta. "We should talk to Luvsan at Mon-Forté Investment Company and Ronnie Lamb at Mongol Amicale. Both those companies are owned mostly by Mongolians, but each company has American investment money. It makes sense for us to talk to them," she says.

17
Fine fiber processors in UB

Our first appointment is with Ronnie Lamb, who, judging by his enthusiastic response to my phone call, is thrilled to show off his processing plant.

"Oh, my god, I can't breathe," says Alta as we enter the Mongol Amicale processing plant warehouse.

"That is the smell of good fiber," says Ronnie Lamb.

"And dirt," says Alta, gagging.

"You've come at a good time. We are washing camel wool today," says Ronnie with pride. He is tall, especially in Mongolia, with red hair and a fair complexion. With his Scottish demeanor and love of wool, he should be wearing a kilt. He is not just in charge here but enthralled with washing, carding, and weaving fiber.

Camel, sheep, and goat wool are bundled, wrapped in burlap, and stacked in the warehouse.

At one end of the warehouse, workers are opening the bales of wool, splitting them into smaller and more manageable pieces, and dropping them into large vats of water where paddles move the heavy fiber to rotating combs that begin the laborious process of separating and washing the wool. As the huge bundles are unpacked and raw fiber is lifted into the washing vats, dust clouds and floating hair rise, making it impossible to see from one end of the building to the other. "I invented this machinery and built it right here. Isn't it amazing?" says Ronnie. "No other processor can handle the volume that I can."

He is beaming, breathing the air like the crisp, fresh air of autumn blowing from the pine forests of Siberia.

"Amicale is all over the world, you know," he tells me. "We buy here, clean, and ship to America, England, or Scotland for finished processing and weaving. We are the largest cashmere importer in Europe," he says with a flourish. "We move the wool from the washing machine into this set of rakes, which spread the wool out for drying. My special combs remove the coarse hair, leaving the fine, soft fiber. Isn't it beautiful?"

"Do you operate near capacity?" I ask.

"We could almost double our finished product if we had the fiber," says Ronnie.

"All the herders are having a hard time," I say, "losing animals to the dzud. We work with herders who want to know that the market is here, in UB, before they go to the trouble of bringing their goods here."

"We can't guarantee anything," he says, turning cautionary.

"No, but would you be interested in having a standing relationship with a group of herders, like an advance buy agreement or a first option to buy?" I ask.

"Well, I can't guarantee a price, you know. It depends on what is happening worldwide for finished cashmere thread," he says. "Those herders don't know about the international market, but it is what runs this company," he says.

"If we could talk with them about what is fair, that would be a good place to start."

"OK, I'll talk to them, but I have competition. Japanese companies, for example. They are putting their money into processing plants all over Mongolia. I've got to compete."

Ronnie is nearly shouting now, partly over the noise of the machinery but also in a heightened state of agitation. He is facing international competition, and I am asking him to strike a deal with the herders. The Mongolians will not close the border with China, which would help the UB processors but would be impossible to enforce and against fair trade agreements. I only want two people to talk to each other: herders and processors. They have a mutual interest, even a dependency, and at the same time, competing interests in the price of fiber.

"Why should I agree to buy from anyone in advance? If I offer a good price, they will be able to sell. If I don't, they won't sell. Isn't that what it is about, really?" he asks.

We are walking toward the end of the drying and carding machinery where soft light brown, almost yellow, camel fiber is emerging, dry and so light I cannot feel it in my hand.

"Good camel hair is just as exquisite as cashmere," says Ronnie. "I love this stuff. Have you seen a camel shedding? It comes off in clumps as if they have a disease. They look awful, but this wool is so beautiful." Ronnie is in a trance, talking to himself now, admiring what the camel can produce and how he can enhance it to make the finest camel cloth on earth. "My greatest thrill is shipping this to Scotland to have it woven. It is always a marvel when a camel coat comes off the line. I know where each hair comes from."

Ronnie starts to walk toward a floor manager to discuss some aspect of the drying process but turns to me before leaving.

"You bring your herders back here, maybe those who did not attend the Summit, and we'll talk," he says.

As Alta and I walk away from the Mongol Amicale plant, Alta says we must talk to Luvsansambuu Tsetsgee at Mon-Forté. "Ronnie is a good processor, but he has a reputation as a tough bargainer," she says.

"He certainly sounded like that when we started talking about business," I say. "You suggested Luvsan before. Let's find him."

18
Sticker shock for the herders

Ambassador LaPorta said the Mon-Forté Company was started in 1994 with a one-million-dollar investment from the Forte Cashmere Company of Boston. With the Embassy's assistance, the owners are transferring their capital interest entirely into Mongolian ownership. Hence, the Ambassador is interested in seeing the Mongolian Mon-Forté Company succeed.

Alta did find Mr. Luvsan, the Mongolian manager of the Mon-Forté Company in UB. He has traveled to the Gobi to meet herders and is very familiar with the hardships of life in the desert.

"He agrees to accompany us with a group of herders if we bring them to UB," says Alta. We want to bring as many of our pilot herders to UB as possible and invite the families that Luvsan met when he traveled to Omnogovi'. If we can get several herders from one aimag in the Gobi connected to a processor such as Mon-Forté, we can get a marketing arrangement going, similar to what we discussed with Mana and Ronnie Lamb. Luvsan suggests that we take them to the Gobi Cashmere Company so that they can see the entire processing operation, start to finish, including the production of retail cashmere products. Mon-Forté does not make finished products in Mongolia, and we want the herders we bring to UB to see the value of the final products.

A bus pulls up in front of the Gobi Cashmere Company. Alta and Luvsan get off first, followed by several herders whom Luvsan met while on a tour of the Gobi recently. We invited Mon-Forté foreign investment company, a Mongolian company initially financed by American capital, and Mongol Amicale to meet the herders in our project.

Luvsan is enthusiastic about making connections with the herders.

"It is a most interesting experience," Luvsan says. "The herders don't know where their cashmere goes when it leaves the Gobi. I want to show them."

Tömör and many of his family and friends are next off the bus. He is wearing a dark purple de'el with a bright yellow waist sash. He has a black patch over his right eye and a weathered face from years in the desert. He stands tall, much taller than the others, giving him a commanding presence. "We want to see what you do with our cashmere," he says.

"This is the largest processor in UB," says Alta. "Before we see the washing and carding machinery, I want to show you the result. They have a store here where you can see the finished products."

We enter a small, neat retail outlet that resembles a specialty shop in a U.S. shopping center. Tömör and the others wander around, touching and holding up sweaters, socks, scarves, and mittens with curiosity. "How much does this cost?" asks Tömör, holding a medium-sized woman's cashmere sweater.

"Here, it costs about one-hundred twenty-five dollars, but in the U.S., it will sell for about two-hundred fifty U.S. dollars," I say. "Here are some magazine ads I brought from the U.S. so you can see the prices." I pass out ads for cashmere garments from Lands' End, Vogue, and J. Crew that I brought with me from the U.S. "A sweater like the one Tömör is holding uses about 0.7 kg of cashmere. What do you get for a kilogram of cashmere when you sell to a trader in the Gobi?" I ask.

"The price now is between seven and eleven dollars for a kilogram of good fiber," says Tömör.

"You are receiving between two and three percent of the retail value of that garment," I say. "The value of the raw material in a finished product like that should be between five and ten percent of the final value. You should be getting twice as much for your fiber," I say.

"Who gets the rest?" asks Tömör.

"Some goes to the local people who bring the goods to UB, the processor, the weaver, the shipper," I say, "but the price then doubles or even triples at the retail store in the U.S. or Europe. "

It is quiet as the herders, Alta and Luvsan, walk around the store. I have placed Luvsan in an awkward position. His processing company only adds a little to the price of a finished good. "Let me say that the UB processors do not profit much from this product." Luvsan looks at me with a faint smile. "We hope you can bring good quality fiber to UB together so that you have some leverage to negotiate a better price, a fair price, so that you and the Mongolian processors like Luvsan can all do better," I say.

"If you sort the fiber in the field before it comes to UB, it will save time for the processors and be more valuable to them so that you can earn a better price," says Alta. "Now, most of it comes in bulk, unsorted. That means the processor must spend much more time."

"We have not sorted before," says Tömör.

"The old system," says Alta.

"We are not yet capitalists," says Tömör.

"No, but you are good herders," says Alta. "Now, I want to show you the rest of the plant.

Tömör strolls through the plant, scrutinizing the machinery.

Figure 21 Cashmere yarn-spinning plan in UB

The workers are just as curious about us as the herders are about the workers, mostly young women whose fingers are deftly flying around spinning machines. Fluffy fiber is coming out of dyeing vats and drying tables to be meticulously spun into thread. The workers carefully monitor the tension on the thread, each person watching several spools as the yarn is spun in a hundreds-long room with dozens of workers. In the next room, much more complex machines take different yarn colors from dozens of spools that feed each weaving machine. Each strand is hooked to tiny sockets that feed the weaving of a sock or sweater, slowly taking shape in front of each worker. Frequently, a worker stops their machine to make a correction, adjust the tension on the thread, or fix a broken thread. Then, the worker starts the machine again. The concentration and intricate manual dexterity required for this job are extraordinary. At the end of the plant, piles of woven garments are inspected. Inspectors, snip, trim, label, fold, and box them, while hundreds of spools of yarn are boxed for shipment to processing plants elsewhere.

"They are so skilled," says Tömör, turning to Luvsan. "These machines…" he says, not finishing the thought, turning to watch. He looks at Luvsan again. "If you pay us for fiber in advance, then we will guarantee delivery," he says.

"Well, there is an idea," says Luvsan, surprised to be amid a negotiating session so quickly. "Is it an advance, an investment, or a loan?" he asks.

"An advance! Otherwise, we have to pay interest," says Tömör.

"You are making a promise to deliver," says Luvsan, thinking how to respond.

"And you are getting our best fiber at a guaranteed price. We come to you first, not to other processors."

"What if the dzud starves and freezes all of your animals, and you cannot bring the fiber that I expect?" says Luvsan.

"We owe it to you as soon as we produce it," says Tömör. "We need to trust each other."

19
Anton's Bistro

In the office the next day, I was thinking about connections between herders and UB processors, but I realized that a deadline was coming soon for the column I was writing for Gobi Business News. This week, we will discuss the subject of goat milk and immunology. On our last visit to the Gobi, Alta and I noticed that herders sometimes bring newborn kids into the ger to keep them warm, inadvertently separating them from the mother. They feed them powdered milk, depriving the young goats or sheep of critical antibodies in their mother's milk. My column is on the importance of keeping newborns with their mothers. I added a pitch about deworming vaccines on the advice of my professional consultant, a veterinarian volunteering to work on animal health in UB. She is a great resource. I finish a draft and look for the editor of Gobi Business News, Layton Croft, who is in an editing and layout room working on the next issue.

Layton and a few other staff members are hovering over proofs. "Good," he says as I hand him my draft. "We were looking for that. Less than five hundred words, I hope. That's all we have room for."

"Should be about right," I say, not knowing for sure.

"You might be interested in our story on mining in the Gobi, around where you and Alta were looking for hay. It looks like the mining companies want grazing land. Stuff for a good story someday," he says.

I notice Vance walking through the office, so I grab a draft of the immunology column and an outline of the overgrazing column and follow him into his office. He is OK with the immunology article, but we have discussed the overgrazing issue for some time.

"So, you want to write about overgrazing. Our animal improvement program is to improve cashmere quality by using elite bucks and removing weak animals from the herd. The consequence will be starting with a smaller herd, but our position with our pilot herders is fiber improvement. We are not directly addressing the problem of overgrazing. I don't think we are ready to put that issue into print in Mongolian just yet," says Vance.

"But if we know that is a problem, why not say it?" I ask.

"That is a different issue than what we agreed to work on. Our program calls for improving genetics, not reducing herd size, though that may be the ultimate result. If you want to write about it, do it in the United Nations Development Program Bulletin, but not in the Gobi Business News. The herders are very sensitive about reducing herd size per se. We need to approach that subject with the herders carefully," he says.

Though we disagree on various approaches to improving life for herders, I agree to write my article on overgrazing for the U.N.D.P. Bulletin, and we adjourn for lunch at Vance's favorite spot, a French restaurant owned by Anton and his wife that caters to expats. We exchange greetings, sitting at a small round café table. "We are here for pleasure, not business," announces Vance, to no avail. Anton is in a business mood.

"Have you heard of Robert Friedland, a world-renowned pros-pector? He is sending notices that he has discovered substan-tial gold and uranium deposits in the South Gobi. His company is Ivanhoe Mines Mongolia, Inc., if you can believe that name," says Anton.

"Pretty graphic," says Vance.

"He always has a stock pick," Vance almost whispers to me.

"The world presents endless opportunities; it is just a matter of which ones we pursue and why," says Vance in French, adding a philosophical air to the conversation while showing little genu-ine interest in Anton's gold fixation.

"We are here for wine, not gold," says Vance. "What have you, Anton?"

Julie, Anton's wife, brings a new French wine, bread, and cheese.

"You are a lucky man, Anton," says Vance. "Steve and I are here alone," referring to me, "or will that change for you, Mr. Parliament?"

"Either Laurie will come here to visit this summer, or I will return to the U.S. It is difficult being so far away and alone," I say, the first time I publicly admit that I am considering returning to the U.S. In his typical response to such interpersonal situations, Vance remains silent.

"And Mr. Vance, have you found a new love yet?" asks Anton.

"Only a friend, in the same building as our offices, but nothing serious," says Vance, admitting more than I had known.

The conversation drifts in and out of rare minerals, new people in town, news from Europe and the U.S., and how hard it is to get good wine here. It is late afternoon, so I walk home through Sukhbaatar Square, the center of UB, where the Great Hall of the People is a concert theater surrounded by many government buildings. The buildings are grey gloomy, two-story stone edifices encircling the vast open square. The fading sunlight is at a sharp angle in winter, so the buildings, though not tall, cast long shadows on the square. The air is hazy from coal smoke, the temperature is below zero, and the streets are busy.

A few hawkers offer to take my picture in front of the Great Hall of the People with a Polaroid camera. Their clientele looks to be Mongolians, probably from the countryside visiting UB and needing evidence of the trip. In the lives of most rural people, a trip to UB is rare. Towards the center of the square, three men in de'els and boots have horses tied to a lamp post. They are talking and passing out a flier. Many others have gathered to listen to them. As I approach, one of the men hands me a flier. Alta translates it as losing grazing land to the mining companies and are being driven out of the Gobi. They have no livelihood and need to keep the land for grazing.

I press my right hand and arm against my chest and then extend them to them as a sign of understanding. I offer my snuff bottle, say goodbye with a Mongolian word that I am told means farewell and a safe journey, and leave, wishing I could do more.

A busy thoroughfare borders the west end of the square. Office workers are starting the trek home. The public bus system is decrepit and unpredictable, so private mini-vans battle the buses for customers. Every bus stop is full of these twelve-passenger mini-vans from which owners have removed the seats to make more room. As vans pull into bus stops, a "jammer" or barker yells out the part of town where the van is going. These are not scheduled routes. The van goes to one or two neighborhoods and drops people close to their homes for a charge. The vans are drastically overcrowded because the public buses may not come at all. However, when they arrive at the bus stops, they drive into the vans, pushing them out because the van business is illegal. The van drivers and jammers pushing more people into their vans are screaming at the bus drivers, who are screaming back, while people are trying to get on one or the other without being run over. In so many other developing countries where I have traveled, one thing you can count on is a public bus, however crowded and unsafe. Even that seems a luxury here. I am glad that I live within walking distance.

A few blocks further are hotels, shops, restaurants, and a city park. The sidewalks are full of people with large fur hats and scarves covering their faces as they puff along. Everyone's breath is visible in the sub-zero, calm, smoky air. In the cold determination of purposeful walking, I usually don't make eye contact with anyone, and no one looks at me.

"Hello, friend," blurts a smiling young woman standing before me. I nearly run over her. "Company?" she asks.

"What?"

"You like company tonight?"

"Oh, thanks, but not tonight," I say and hustle on.

Alta tells me later that they are the children of herders who cannot afford to keep their daughters in the countryside if they are not married. They come to the city, but commercial sex work is often the only job they can get. The average age of the UB population is about 25 years, and it is getting younger, says Alta. "Those young girls," says Alta. "What to do?"

"I hope that Batsukh can keep her daughters safe," I tell Alta.

"They need to work, but they all cannot herd," says Alta. "I heard that the volunteer group downstairs in our office, you know, where the veterinarian you work with is doing something with small businesses in the countryside, possibly a way for women to earn money."

"Do you think this could become part of our program?" I ask

"Of course," says Alta. "It would give women in a herding family something to do so they don't have to come to UB looking for work and finding none."

"I think you have many opportunities that you can follow, Alta, and I also want you to know that I recently discussed with Vance the possibility of my returning to the U.S. fairly soon. Most of my reasons are personal," I say.

"We have just started," she says.

"I know, and there is a very long way to go. Leaving is tough, but you and others here know the people. I hope this way of life survives, but you have a good idea of what to do. Just promise me that if I return five years from now, Mongolia will not be overrun by dairy farms and strip mines."

"We will always have cashmere," she says.

20
Emerging markets: Price and exchange in the Gobi Desert

The herders who live closer to the cities, such as UB or Dalanzadgad, have more information on marketing and more possibilities to sell their products. They can get higher prices. If they are remote, selling their products in the market is very difficult, so they produce dairy goods and use them independently. Now, meat prices are going up. Meat is costly in UB and other cities. The herders sell cheaply, and there are middlemen, the changers, who profit from meat and dairy products. The milk trade depends on the season. In winter and spring, the milk price is good enough for the herders, but in the summertime, the milk price is very low because everyone is producing milk then.

We have talked about cooperatives for many years, starting in 1996, but their success is not good enough. First, the herders always move to find grass, so they cannot work together. Second is low education because it is difficult for herder families to attend school. They don't understand the advantage of cooperating. Also, they don't like to keep records of how they are doing and how much they have produced. They don't have time to maintain records and minutes of meetings.

Trust among the herders is also essential. From ancient times, the herders moved around separately, so trust among herder families is very difficult. They move away from each other as they migrate. Leadership among the herders is also critical. If there are good leaders, then the cooperatives do well. I think we have about 50 or 60 successful cooperatives in the country. All the cooperatives have good leaders who can manage the herders. They can build trust among the herders.

According to Mongolian cooperative law you need nine family members, and it isn't easy to get them and keep them together. It is especially important to find young leaders. Most of the herders are getting old, while the young people are going to the cities or universities.

There are young herders, but they are poor. In the 1990s, herding was privatized, and many young boys went to look for their herds to take care of their animals. But they left school, so their education is lower. Management skills are very important for younger herders. Some donor organizations are working with the herders, but it is not enough. There is no technical support from the government side.

First, we must build herder groups with three or four herder households. If they can cooperate successfully, they can build or expand the cooperative based on the family business. If they have members from the same family, like daughters and sons, working with parents, they can cooperate more successfully. They need to get some profits from the cooperative when it is run like a business; then it will work. If all members are encouraged to take professional responsibility, they will gain each other's trust. It takes time.

Bayarmaa Chimedtseren also worked for nine years for the Gobi Initiative and is now the Deputy Director for the Peri-Urban Rangeland project. We organize meetings, trade fairs, and market days in each aimag. These events were quite successful and continue, with many trade fairs and market days in UB, especially in the fall, which is called Fall Green Days. People bring many products, especially dairy products.

We focused on linking cashmere with international markets but do not always meet the quality requirements of global traders. We collect fine cashmere from many aimags and send it to the laboratory to test the micron. Some herders complained that the lab tests were not accurate. What is happening is that the changers are buying cashmere and reselling it, taking the profits.

There is a market network for meat established in some regional centers. Last year, the Chinese bought meat, so Mongolians sold lots of meat to Chinese traders, and the price was high. Goats were mostly sold as whole carcasses. We think the Chinese might buy carcass meat, trim it, and sell it back to Mongolians at a higher price.

The Mongolian government buys meat for reserves to sell when the meat price is too high. However, the government probably could not reserve enough meat to affect the price. People are complaining that Mongolia is a meat-producing country where the meat price is very high. The Mongolian government needs to take action to keep the price reasonable.

Milk is comparatively available. Mongolia does not export milk, so it can be purchased here in the country. Milk prices are stagnant.

Regarding wool, the government approved a program for herders who sell wool to domestic factories. The herders receive a subsidy, which is a good incentive for them to increase their production of cashmere and camel fiber and sell it to UB processors. This subsidy is only available for cooperative herders' organizations, not for individual herders. This policy also encourages good pasture management because the cooperatives sustainably maintain the number of families and animals.

The formation of cooperatives and herder groups is beneficial, but the changers buy fiber and take a portion of the price. The changers are very active and know how to operate, even illegally.

With the formation of cooperatives, agricultural product prices have increased, positively impacting herders' livelihoods.

There are around 117,000 herder households in Mongolia. We will keep working with them however we can. Cooperation is the only solution.

21
Visit to Dalanzadgad

As I prepare to leave Mangolia, Alta is talking with a farmer from Minnesota named David, who is volunteering time in Mongolia as a business development consultant. He is a dairy farmer who helped a large dairy cooperative called Land-O-Lakes create new dairy products in the United States. His idea is to do the same thing in Mongolia using goat, sheep, and camel milk. David has been traveling to Dalanzadgad, an aimag center in the South Gobi, to work with herders on new product development. Alta and David are setting up an event called Cashmere Market Days to demonstrate some new goat and sheep milk products and to bring herders and processors together, like a trade fair in fiber.

I want to see the marketing days in Dalanzadgad, so I will stop there before leaving Mongolia for Beijing, China.

Dalanzadgad is a small and unassuming town. As you enter, the first structures you see are the rusted steel frames of old warehouses, followed by rows of vertical slat fencing surrounding gers. Trails enter the town from all directions and merge into wide gravel paths. The town center has one definable roadway lined by commercial and retail buildings, with the government center at the end. Cashmere Market Days are being held in an open area surrounding the commercial buildings.

The market is full of horses, jeeps, and old military trucks. Bales of fiber from camel, sheep, and goat are being unloaded into the warehouses and onto tables in the market area. The herders' families begin sorting coarse hair from fine, almost strand by strand, by breaking open the bales of mixed fiber and spreading them on tables. The processors observe the sorting and grading to confirm the quality. David has arranged for technicians with sophisticated equipment to conduct micron measurements to determine the thickness of a strand of selected hair and confirm its value.

The organizers estimate that one-hundred fifty tons of cashmere from individual herders and herder cooperatives will change hands today, with prices from twenty to twenty-five percent higher than what a herder would get selling raw fiber to a processing plant in UB. Having more than one buyer is an advantage to the seller because of competitive pricing, and having many herders with high-quality fiber is an advantage to the buyer because they can look for the finest fiber available.

Food stalls sell hot pots, fried meat, and brick tea in one corner of the market. Packages of goat cheese, yogurt, and dried milk produced by women herders are on display. A woman is selling dried meat next to the stand with goat cheese.

"What kind of meat is this?" I ask.

"Dried mutton, but I have some dried beef also," she says, holding up some dried meat in a clear plastic vacuum pack with Russian printing. I look at the packaging more closely. On the back is some English stating where the meat is from.

"Alta, look at this," I exclaim. "This dried beef is from Minong, Wisconsin, a place called Jack Link's Beef Jerky. That is just a few miles from where Laurie lives in northern Wisconsin. Can you imagine that they packaged and printed this dried beef in Wisconsin and sent it to Russia? And here it is for sale in a market in the Gobi Desert."

"One thing Mongolians don't need from the States is dried meat," she says. "Why is it here?"

"Foreign aid, probably," I say. "If we bought Mongolians a vacuum packaging machine, they could make dried meat themselves."

"Is it in the budget?" she asks.

"See to it next year," I say.

We walk through the market until we run into the mayor and other government officials. Alta introduces me. They are all very proud of this event and pleased that it is being held in Dalanzadgad.

The mayor wants me to try something, so he guides me to a table with small cardboard containers, like milk cartons. He picks up a container, opens it, and offers it to me.

I am sure that my eyes open widely as I smile and sample. It is the most magnificent, luscious, and richest milk I have ever tasted. "This is incredible," I say. "What is it?"

"Camel milk," he says gleefully.

"Amazing. It is richer than pure cream."

"We are starting to package it here," he says. "A new business, the first time anyone has ever done it."

"The whole family can work on these products," says Alta, "and ship them to UB."

"Or to China and then to the United States," I say, which is the route I am about to take. Alta, the other staff members here, and I have lunch together for my last Mongolian hot pot soup. They give me a painting of a horseman, reminding me of Chuulunbaatar in his competitive riding days, and a small hand-woven wool rug, which I carry with me.

"I will not say goodbye because I will see you again in the States," says Alta.

"And I will see you again here, someday. You have so much to do, and I know you can do it," I say.

It is very difficult to leave this wonderful group of people, with their sharp sense of humor and deep dedication to the country. As I leave, no more words are needed.

From Dalanzadgad, I go east to Sainshand in Dorngovi aimag, a small sum center on the train route between UB and Beijing. The train continues from Beijing to Tianjin, the major port for overseas goods. The train tracks in Mongolia and throughout Russia are a different gauge than those in China. The train from UB stops at the border, and the train cars are lifted from one set of bogie wheels and placed down on another sized set. Then they move on. How can the herders connect with the outside world when this is the only route out? I am thinking about Chuulunbaatar and Udval, their son Batzorig, their daughter-in-law Naran, and their other sons; about Batsukh and her eighteen children; Tömör with a patch on his left eye and his one-eyed penetrating stare; and Dembereldorj who has the most animals and the largest stack of frozen carcasses in the Gobi. How can they trade on the world market when they cannot reach it?

Beihai Park is in the center of Beijing, northwest of the Forbidden City and Tiananmen Square. Before the Forbidden City, this park was the location of Kublai Khan's palace. All that remains of the Mongolian presence is a jar made of green jade near the southern entrance to the park. An information sheet notes that this park was Madame Mao's favorite spot for meditation. The notes do not mention the Mongolians.

The park is an island surrounded by lakes. I sit at an outdoor café admiring the beauty of the White Pagoda through the Yong'an Temple, built in 1651 for a visit by the Dalai Lama. A man with a bucket and a large "bi" or Chinese lettering brush stands on the sidewalk across from where I sit. He dips his brush into the bucket and paints Chinese characters on the sidewalk. His audacity strikes me presuming he can paint the sidewalk without being arrested. I am intrigued by his deliberate and meticulous writing.

The sidewalk is sectioned into two-foot squares like tile. The painter selects a quite large area. He begins in the upper right-hand corner of the section of sidewalk he has chosen, one large character for each square, writing downward and then moving to the left as he finishes a column, as in classical Chinese writing.

People walk by, stop, and read what he is writing. They do not speak with the poet, nor does he acknowledge them. I wish that I could read what he is writing.

He stops to dip his brush, looks over his work, and continues writing.

My curiosity overwhelms me. I walk to where he is working without bothering him, hoping that I can find someone who can translate. The people who walk by read in silence. It is not appropriate to interrupt them. An older Chinese man stops next to me.

"What?" I ask in my rudimentary Chinese, pointing to the writing.

"A poem," the man says in English.

Figure 22 Sidewalk poet writing with water

After half an hour, I notice that the Chinese characters in the upper right-hand corner of the sidewalk where he started are fading; they are disappearing. I approach the writer.

"What are you using?" I ask, pointing to his paint can.

"Water," he says.

Nothing is permanent.

I leave Beihai Park, the Forbidden City, Peking, Mongolia, and Central Asia, bringing it all with me, never to forget.

Afterwords

Update 2024. The climate challenge is getting more difficult. Mongolia suffered a devastating dzud in 2023–2024 with the most snow in 49 years, resulting in the death of 5.9 million livestock, and about 60 million facing starvation. The livelihood of the herding lifestyle itself is under threat.[1]

The United Nations Office of the Coordination of Humanitarian Affairs issued the following pleas for help. "Despite their diligent preparedness actions, herders require our unwavering support to mitigate risks. Thus, I earnestly appeal for international assistance and the mobilization of additional resources to prevent a further deterioration in the humanitarian situation in Mongolia." Tapan Mischra, United Nations Resident Coordinator in Mongolia.[2]

Without cashmere herding families, the entire culture may not survive.

Notes

Preface/Introduction

1. "On the Cashmere Trail," *Mongolia Survey*, Issue No.7, The Mongolia Society, Fall 2000.

2. Louisa Lim, "Mongolia in Transition: Mongolians Seek Fortune in Gold, But At a Cost," *National Public Radio*, Washington, D.C., 9 September 2009. Available at: https://www.npr.org/2009/09/07/112516360/mongolians-seek-fortune-in-gold-but-at-a-cost

3. Sarah J. Wachter, "Pastoralism Unraveling in Mongolia," *The New York Times*, 8 December 2009, *Special Report: Business of Green*. Available at: https://www.nytimes.com/2009/12/08/business/global/08iht-rbogcash.html?searchResultPosition=1

4. Jane Perlez and Ismail Khan, "U.S. Aid Worker Slain in Pakistan," *The New York Times*, 13 November 2008.

Afterwords

1. John Yoon, Khaliun Bayartsogt and Somini Sengupta, "A Harsh Mongolian Winter Leaves Millions of Livestock Dead," *The New York Times*, March 29, 2024.

2. Tapan Mischra, U.N. Resident Coordinator in Mongolia, as quoted in Mongolia Dzud Response Plan 2024. https://humanitarianaction.info/plan/1197/article/mongolia-dzud-response-plan-2024

Recommended assignments and discussion questions

1. In your judgment, what are the principal obstacles to the continued livelihood of Mongolian nomadic culture? Can it survive, or are the logistical, environmental, and political obstacles insurmountable for a sustainable society? Or can resourceful Mongolians overcome these obstacles?

2. Mongolia is a landlocked country, as the book describes. That means they have no ports for exporting goods, either raw fiber or finished products. The only way out for trade is either through Russia or China. There are enormous risks in dealing with either country as a partner. However, Mongolia has no choice. If you were advising Mongolian trade officers, which country would be most compatible with Mongolia? What are the risks in forming a trading partnership with the country of your choice? What conditions would you propose for Mongolia to place on any trade agreement with Russia or China?

3. Identify the major natural resources of Mongolia. Don't forget the intangible ones, such as natural beauty. Which resources are not fully developed? Is the country overly reliant on one product? Outline a development plan emphasizing increased diversity.

4.   According to the Mongolian Constitution, all land is owned
     and controlled by the central government on behalf of the
     general population. This is a statutory reflection of the need
     to move animals around for grassland. Does this agricul-
     tural requirement for sheep and goats preclude the devel-
     opment of other forms of livestock or agricultural land use
     that might be more profitable in the long run? Additionaly,
     is private land ownership a necessary condition for building
     a successful and competitive economy?

References

People I worked with while in Mongolia or interviewed for this book.

Alphonse F. LaPorta	U.S. Ambassador to Mongolia
Angha	Botzorig's wife, Chuulunbaatar's daughter-in-law
Anton	Owner of a bistro in UB
Ariuna	Staff, Gobi Initiative Animal Improvement Program
Banzragchyn Altantsetseg (Alta)	Program Officer for Agriculture, Gobi Initiative; Director of Peri-Urban program, Millennium Challenge Account
Batzorig	Chuulunbaatar's first son
Bayila/Baatar	Herders in China, brothers
Buyan	Translator
Chin-Erdene	Mongolian Foreign Investment and Trade Agency
Christopher Finch	Executive Director, Mongolian Foundation for Open Society (SOROS Foundation)

Chuulunbaatar	Pilot herder in the agricultural improvement program
Chuluunbatyn Yondonnorov	Governor, Uvurkhangai aimag, Burd Sum
Davaasambuu	Government of Mongolia, External Relations
Debra Rasmussen	Project Manager, Agriteam Canada
Delgerjargal Uvsh (Degi)	Raised in UB; a doctoral student in political science at the University of Wisconsin, Madison
Delgermend Tserendamba (Degi)	Graduate of the University of Minnesota with field work on women herders; now at the Mongolian National Psychological Association in UB
Dembereldorj	Herder
Edward Burgells	Country Director, United States Agency for International Development, Mongolia
Erdene	President, Mongolian Association of Private Herder's Cooperatives, and Executive Director, Training Center of Cooperatives, UB
Gordon Winward	Kiwi Milk Production, Aukland, New Zealand
Hiroshi Fujimoto	Japanese Embassy, Third Secretary
Itgel	Business development staff, Gobi Initiative, and economic development adviser, the Asian Development Bank, UB
Jargalsaikhan	General Director, Buyan Holding Company, UB

Kangbin Zheng	Adviser, The World Bank, partnership group
Kay Coombs	H.M. Ambassador, British Embassy Ulaanbaatar
Keith Bradsher	New York Times station chief, Hong Kong
Kenta Goto	Associate expert, governance and economic transition, United Nations Development Program, UB
Kiyoshi Hara	Japan International Cooperative Association in Mongolia, Ministry of Agriculture and Industry, Tokyo
Layton Croft	Editor, Gobi Business News, Executive Vice President for Corporate Affairs, Ivanhoe Mines Mongolia, Inc., Mongolia
Luvsandorj	Executive Director, Mongol Amicale, Mongolian American Joint Venture, UB
Luvsansambuu Tsetsgee	Administrative Director, Mon-Forté Foreign Investment Company in Mongolia
Mana	Processing plant owner, UB
Maidar	Founder, Green Revolution Mongolia and promoter of technological entrepreneurial incubators
Michael G. Martin	Resident representative to Mongolia, International Monetary Fund
Nadmid	National Association of Mongolian Agricultural Cooperatives
Naran	Driver
Nergui	Herder turned miner

Udval	Chuulunbaatar's wife
Oyuna, Oidov. Oyuntsetseg	Women's Information and Research Centre, and the Gender Center for Sustainable Development, UB
Oyun, S., Dr. Sanjaasurengiin	Member of Parliament, Minister of Nature and Green Development; Minister of Foreign Affairs; leader of Democratic Party
Robert Friedland	Owner, previous CEO Ivanhoe Mines Mongolia, Inc.
Ronnie Lamb	President, Mongol Amicale
Sanjaasürengiin Zorig	Nominated for Prime Minister by the Democratic Party; brother of Dr. S. Oyun, MP; deceased.
Sarangerel	Boot-maker
Satoshi Matoba	Japanese Embassy to Mongolia, Second Secretary for Economic Cooperation
Shinzo Tanaka	Trade and Development Bank of Mongolia
Sodnomyn Chinzorig	Chairman of the Presidium of Citizen's Representatives of Uburkhangai aimag
Spilhaus, Karl	President, Cashmere and Camel Hair Manufacturers Institute, Boston, MA
Stephen D. Vance (deceased)	Chief of Party, Gobi Initiative, Mercy Corps International; field staff Foundation for Cooperative Housing International, assassinated by extremists in Pashawar, Pakistan while on assignment.
Tömör	Herder

Tsendmaa	Executive Director, Mongolian Wool and Cashmere Federation
Tserenchunt Legden	Raised in a herding family, and now Senior lecturer of Mongolian language, Indiana University; on the board of The Mongolia Society; mother of Delgerjargal (Degi) Uvsh
William Siemering	Founder, national public radio and Special Advisor, Open Society Institute, Mongolia

Bibliographic references

Batbayar, Tsedendambyn. 2002. *Modern Mongolia: A Concise History*. Ulaanbaatar: Mongolian Center for Scientific and Technological Information. This book provides a brief yet useful synopsis of Mongolia's history, beginning with its 11th-century expansionist experience.

Chatwin, Bruce. 1997. *Anatomy of Restlessness: Selected Writings 1969–1989*. Edited by Jan Borm. New York: Penguin Books. This collection of Chatwin's essays includes *The Nomadic Alternative* and *It's a Nomad Nomad World*, which explore themes of wandering and restlessness. Chatwin also offers observations on life in Argentina (*In Patagonia*) and the Aboriginal Dreaming Tracks in *The Songlines* (New York: Viking Penguin, 1987).

Fijn, Natasha. 2011. *Living with Herds: Human-Animal Coexistence in Mongolia*. New York and Cambridge: Cambridge University Press. A thorough and compelling study offering personal observations of human-animal interaction in Mongolia from the perspective of an anthropologist and filmmaker.

Global Policy Forum. 2005. "Mongolian Woman MP Takes Aim at Corruption." *Radio Free Asia*, April 25. Retrieved October 14, 2009. globalpolicy.org/component/content/article/172/30287.html.

Greene, Joshua. 2013. *Moral Tribes: Emotion, Reason, and the Gap Between Us and Them*. New York: The Penguin Press. A discussion on moral guidance and altruism in addressing the tragedy of the commons.

Hardin, Garrett. 1968. "The Tragedy of the Commons." *Science* 112, no. 3859: 1243–1248. DOI: 10.1126/science.162.3859.1243.

A widely cited article analyzing individual interests in a communal setting where resources must be shared.

Hůlová, Petra. 2009. *All This Belongs to Me: A Novel*. Evanston, IL: Northwestern University Press. A novel told from the perspectives of female family members living in Mongolia, depicting the challenges of transitioning from nomadic to urban life.

Humphrey, Caroline, and David Andrews Sneath. 1999. *The End of Nomadism? Society, State, and the Environment in Inner Asia*. Durham, NC: Duke University Press. A comparative study of nomadic communities undergoing significant social, political, environmental, and economic changes.

Ingold, Tim. 2000. *The Perception of the Environment: Essays on Livelihood, Dwelling, and Skill*. London and New York: Routledge (ebook 2002). A collection of essays by a social anthropologist on human-animal coexistence and adaptation in challenging environments.

Knauft, Bruce M., Richard Taupier (Eds.), and Lkham Purevjav (Managing Ed.). 2012. *Mongolians After Socialism: Politics, Economy, Religion*. Ulaanbaatar, Mongolia: Admon Press. A collection of articles examining governance, the economy, wealth disparity, contemporary religion, and cultural history in post-socialist Mongolia.

Le Clézio, J. M. G. 2009. *Desert*. Translated by C. Dickson. Boston: A Verba Mundi Book, David R. Godine. Originally published in French as *Désert* (Paris: Editions Gallimard, 1980). A novel about the "last free men," nomads of the Moroccan desert facing cultural extinction before World War I.

Morozova, Ly. 2010. "Political Parties Against the Background of Neo-Liberal Reform in Present-Day Mongolia." *Mongolian Studies: Journal of the Mongolia Society* 32: 61–84. Bloomington, IN: Indiana University, The Mongolia Society. A political analysis of Mongolia's transition to market economics.

Ostrom, Elinor. 1990. *Governing the Commons: The Evolution of Institutions of Collective Action.* Cambridge and New York: Press Syndicate of the University of Cambridge. A seminal work by the Nobel Laureate in Economics on managing common property resources through cooperative enterprise.

Paley, Matthieu (photographs) and Mareile Paley (text). 2010. *Mongolie: La Route de l'Horizon.* Paris: Éditions de La Martinière. A beautifully illustrated photographic account of Mongolia's people and landscapes.

Parliament, Stephen. 2000. "Cashmere Grading and Marketing." *Gobi Business News*, March. Articles based on fieldwork in the Gobi Desert as Program Director for Agriculture, Gobi Initiative, Mercy Corps International, under contract with USAID.

—— "Post-Partum Care and Disease Management." *Gobi Business News*, April.

—— "Effects of the Dzud Disaster on the Cashmere Market for Herders." *United Nations Development Program, Mongolia Bulletin*, April.

—— *Guide to Husbandry and Marketing for Nomadic Herders.* Gobi Regional Economic Growth Initiative and USAID, March.

—— "Land Management and Overgrazing, Mongolia and the Dzud." *Newsletter, The Mongolian Society, Indiana University*, May.

—— "On the Cashmere Trail." *Mongolia Survey*, no. 7. The Mongolia Society, Fall.

Riouall, Gaëlle. 2006. *Chez les Nomades de Mongolie.* Paris: L'Harmattan. A detailed description of the daily lives of Mongolian nomads, with drawings, photographs, and diagrams.

Rossabi, Morris. 2005. *Modern Mongolia: From Khans to Commissars to Capitalists.* Berkeley and Los Angeles: University of California Press. A political history of Mongolia's transition from a Soviet satellite state to an independent, market-based democracy.

Sills, Linda. 1999. Information on Batsuh is partially taken from Sills' report in Ulaanbaatar, "Mongolia: Women Redefine Their Roles." *BBC News, World: Asia Pacific*, November 24, 14:03 GMT. Retrieved October 14, 2009. news.bbc.co.uk/2/hi/asiapacific/533527.stm.

Terzani, Tiziano. 1997. *The Fortune-Teller Told Me: Earthbound Travels in the Far East*. New York: Three Rivers Press. A travelogue spanning Singapore, China, Mongolia, and Europe, featuring encounters with fascinating people in Ulaanbaatar.

Tsedev, Khishigjargal, and S. Tserenbat. 2000. *Magnificent Cashmere: A Look Into the Luxurious Clothing Fiber of Mongolia*. Dodge City, KS: High Plains Publishers, Inc. A technical and visual exploration of fine cashmere fiber, comparing it with mohair, camel, llama, alpaca, guanaco, and vicuña.

Tschinag, Galsan. 2007. *The Blue Sky*. Translated from the German by Katharina Rout. Minneapolis, MN: Milkweed Editions. A novel about a young Tuvan nomadic herder living in Mongolia's Altai Mountains, depicting their rugged way of life. Originally published by Suhrkamp Verlag, Frankfurt am Main, 1994.

—— 2010. *The Gray Earth*. Minneapolis, MN: Milkweed Editions. The second book in a trilogy that begins with *The Blue Sky*.

Weatherford, Jack. 2005. *Genghis Khan and the Making of the Modern World*. New York: Three Rivers Press. A historical account of Mongolian influence on modern civilization, spanning from China to Western Europe.

Whewell, Tim. 2000. "Women Steppe Out in Mongolia." *BBC News, Crossing Continents: Asia*, July, 16:57 GMT. news.bbc.co.uk/2/hi/programmes/crossing_continents/asia/854144.stm.

Recommended further reading or watching

Fijn, Natasha. 2011. *Living with Herds: Human-Animal Coexistence in Mongolia*. New York and Cambridge: Cambridge University Press.

Humphrey, Caroline, and David Andrews Sneath. 1999. *The End of Nomadism? Society, State, and the Environment in Inner Asia*. Durham, NC: Duke University Press.

Ostrom, Elinor. 1990. *Governing the Commons: The Evolution of Institutions of Collective Action*. Cambridge and New York: Press Syndicate of the University of Cambridge.

Tschinag, Galsan. 2007. *The Blue Sky*. Translated from the German by Katharina Rout. Minneapolis, MN: Milkweed Editions.

Weatherford, Jack. 2005. *Genghis Khan and the Making of the Modern World*. New York: Three Rivers Press.

Film

The Story of the Weeping Camel. 2003. Directed by Byambasuren Davaa and Luigi Falorni. Germany: ThinkFilm. https://archive.org/details/TheStoryOfTheWeepingCamel.

Wild Horses of Mongolia with Julia Roberts. 2000. Directed by Nigel Cole. USA: PBS. https://www.pbs.org/wnet/nature/wild-horses-of-mongolia-with-julia-roberts-introduction/2887/.

Wild Horses of Mangolia with Julia Roberts

Index

www.ingramcontent.com/pod-product-compliance
Lightning Source LLC
Chambersburg PA
CBHW061250220326
41599CB00028B/5594